ORGANIZATION DEVELOPMENT

ORGANIZATION DEVELOPMENT

Developing the processes and resources for high-tech businesses

By

BAISHAM CHATTERJEE

iUniverse, Inc.
Bloomington

Organization Development
Developing the processes and resources for high-tech businesses

iUniverse books may be ordered through booksellers or by contacting:

iUniverse
1663 Liberty Drive
Bloomington, IN 47403
www.iuniverse.com
1-800-Authors (1-800-288-4677)

ISBN: 978-1-4620-2129-1 (sc)
ISBN: 978-1-4620-2130-7 (ebk)

Printed in the United States of America

iUniverse rev. date: 05/10/2011

CONTENTS

AUTHOR BIOGRAPHY

The author completed his MBA from a very reputed institute in India around 4 years back. He has completed a research assistantship at UNBSJ. He has written 2 books, **Reconstructing competition and its processes** and **Finding the Innovation Gap** both at iUniverse. The author has created many blogs that have high potentiality in consultancy generating through them very creative ideas and interests in the western economy. The blogs were immense and have created a high impact in e-learning both in strategy and innovation. The author has completed many projects and have participated and contributed and completed projects of more than hundred pages mainly the project at PANalytical. The author has got around 37 research findings that have created immense potentiality for further development in the North American economy. The author has high perception skills and thoughts in the field on marketing, business strategy, e-commerce, innovation and organization development where he has proved.

ACKNOWLEDGEMENTS

The author wants to thank Gregory J Fleet (UNBSJ), Stephen Bernhut (Ivey business school)and Mark Hollingworth (McGill University). I also thank my little brother Ganesh for the moral support I received from him.

I also want to thank Paul Mortimer of the UNB Ward Chipman library

I also thank Nirmal Chakrabarti (Royal Philips Electronics India) and Manik Mondal (NIT Durgapur library) for the support they have given to me.

PREFACE & INTRODUCTION

The ideas in the different chapters and their components have been described.

In chapter 1 on goals the ideas behind forming long term goals, communication system and sources of motivation and organization myopia have been discussed. Other ideas discussed are on how goals are set, goal paths, stretch goals, CRM, corporate goals have also been discussed. Moreover social and environmental models, competence levels, goals for product prototype strategy, goals for pursuing investment strategy and how an R&D can set goals can also be discussed.

In chapter 2 on vision that helps create environmental dynamism and hostility, vision helps the strategists explore the quality domain. A lot of discussions have also been made on the 360 degree feedback and what impact visionary leadership has on organizations and further discussions have also been made on the visionary model, theory and influence of a visionary. Questions arise on how visionary goals can be created, ideas about creating visionary companies have also been discussed. A lot of discussions have been made on goal internalization and how innovation, goals and vision are interrelated.

Chapter 3 on incentives discusses a lot on compensation systems and monetary incentives. A clear picture has also been formed on the WOM incentive program and incentives in e-governance and e-democracy and ISSON (incentive for secret sharing in self organized networks). A relationship has been brought between innovation, reward and recognition program and investment incentive. There are also certain discussions about compensation program, performance based incentive and management buyout. Social incentives has also been given some importance as well as incentives in retail and dealership. All these information covers in depth the third chapter.

Chapter 4 on best practices and balanced scorecard consists of ideas like JIT, TQM and other processes and similarly other areas of study like TQM is also a best practice. Moreover the two types of benchmarking like co-operative benchmarking and competitive benchmarking has been referred. A 4 stage description has also been made on the leadership driven performance best practice. Product flexibility and internal integration across functional boundaries has also been described. There are other types of benchmarking activities like flexibility that are related to design, mix and volume. A value generation strategy and quality improvement strategy has also been described.

In chapter 5 on new and breakthrough technology it has been described that new technology often depends on R&D and competitive strengths. In the same scenario evolution of virtual businesses, creation of e-marketplaces and searching new capabilities has also been described. Production processes, competitive pressures and technological perceptions have also been described. Connections to internal IT infrastructure, push technology and different risk factors have been described.

In chapter 6 on enterprise resource planning it helps in generating real time data, enabling linkages across functional area, it helps in developing processes and understanding environmental information systems, moreover ERP installations and how ERP systems have further developed the scope of standard software. CRM and ECM, CAD, Cam are other system objectives that have been described. The life cycle of the ERP process, BI and many other subpart of ERP have been described.

In chapter 7 on supply chain of ICT and manufacturing firms the described facts are modularity, pre-sourcing and relationship and development with suppliers has been described. Very recent developments are on OEM (original equipment manufacturers), ESP (environmentally sound products) and SGO or strategic green orientation. A brief development has been made on SCM models, SC performance and SC integration. More concepts have been developed on SCM developments of manufacturing and ICT firms. More developments have been made on GIS, EDI, supply chain coordinator and different flexibility modules. It has been concluded by talking of E-supply chains.

In Chapter 8 based on interrelationships and economies of scale description has been formed on the 3 types of interrelationships, controlled decentralization as well as TMT. Importance have been given to building a corporate data model, performance measurement systems, another area of study is the interrelationship between network externalities and information asymmetry, moreover relationship has been created between corporate identity and reputed brand. There is also a description of the rise of economies of scale.

In Chapter 9 on linkage it has been described why linkage is so important. Similarly there is description of various types of linkages like strategic planning and purchasing, moreover Likert scales have been described in detail. Moreover SCM linkage and linkage between production efficiency and adopting new technology, between marketing and balancing of priorities has been defined. A lot of developments have been made on linkage and the value process, channel linkage and how linkages bring competitive advantage and many more developments.

In chapter 10 on performance measurement it is shown that performance measurement has an operational focus and computer integrated manufacturing, flexible manufacturing systems, just in time, optimized production technology and TQM are the different types of performance measurement systems. Analysis has also been done on the main components of performance measurement as well as the core of performance measurement. Descriptions have been made on how performance measurements systems helps in monitoring and maintaining organizational control, moreover it has been said that a performance measurement system should be a dynamic system and more details should be given on logistics performance. Performance gaps and accountability has also been developed on. More developments have been made on business performance measurement and the composition of performance measurement systems. Composition of financial performance and composition of value chain with performance is also to be seen. At the end of the chapter non-financial manufacturing performance measure and the activity based management project has been described.

Chapter 11 describes competitive benchmarking as a typical type of benchmarking that brings success and developments have been made on strategic reference framework, process planning and describes competitive cost benchmarking. The chapter describes how competitive benchmarking can be followed from Deming's ideas and how it brings competitive success. The chapter also describes competitive circumstances, channel power and why competitive benchmarking is also known as environmental benchmarking. Developments on JIT, strategic benchmarking, pre-analysis of benchmark ideas, effective benchmarking and future competences which are all sub-parts of competitive benchmarking.

Chapter 12 on analytical hierarchy process describes firstly why it is a multi-attribute decision tool and describes the process and application. It is said that AHP can access simulation software packages and develops manufacturing priorities. The different power criteria and processes of AHP has been given as well as the different methodologies derived from AHP. The factors for application of AHP and the AHP-PGP approach which is also a decision making approach is described. The chapter ends with AHPs manufacturing outlook and its relationship with suppliers.

Chapter 13 on E-commerce strategies describes the EC business model which is defined as a competitive strategy for the marketplace. Descriptions has been made on EC information systems, profit discrepancies of E-commerce, processes from beginning to end of a product life cycle, what is also very important is the communication platform described in EC, e-servicescape, EDI, ICT diffusion, M-commerce and the ideas involved in building a E-commerce web is discussed. The whole chapter also includes the initial framework.

Chapter 14 on core competence and organizational learning defines core competencies and describes the types of competencies. The components and success factors for studying core competence has been classified. Points like skills, organizational success, joint learning, individual competencies, current competence, empirical capabilities, McKinsey's 7s framework, necessities of core competence, organizational competence, performance based competence, and manufacturing competence and many more points have been described in detail. In organizational learning the types of organizational learning have been described, moreover Taylorism, how a business organization has to be to develop, pressure, organizational performance, technology tools have been developed on. Corporate restructuring, organization memory, capitalization of competences has been stated, organizational framework, organizational culture and development process of planning activities have been described, systems, structural and interpretative models and current organizational foundation has been described.

Chapter 15 on transformation shows that transformation is an integral part of organization development, its various types and forms and on what it depends has been described. Globalization, E-learning, organizational change, using transformation to understand the process of imitation, CRM strategies, how channel strategies affect an organization, transformation forms have also been described in detail. The chapter has concluded by describing work transformations and communication usage.

Chapter 16 on organizational design and internal communications descriptions and definitions has been made on organizational direction, organizational communication, boundary spanners, boundary roles and various new models have been described. Pooled interdependence, organizational size, there are ways of modifying an organization, Carnegie model, communications, the three basic innovation processes, inter-organizational communications, modern innovation change and organizational boundaries have also been described.

CHAPTER 1

Developing the goals of a business
(BY BAISHAM CHATTERJEE)

Long term goals are the key achievement areas of a business. Goals define the country we stay in and create a most probable environment for the future to sustain and grow with. Modern computer simulations have helped NASA create goals and predictions for the next 200 years. As the one year budget cycles or the two to four year political cycles shape government plans and budgets. Goals can be created by compiling many short term goals to form long term goals. The end product of a national long term strategy are short term agendas focused more on linking tactical goals to performance measures. Futurist ideas generally consists of some audacious goals of the past, as well as some failures of imagination and are required mainly in places which require long term possibilities and continuous changes that can radically alter the future. But ideas created from these should be reasonable and achievable and dominated by a group of people who can go beyond what has been assumed and is achievable and can energize people. They motivate and align efforts stimulating forward progress through multiple organizations and drive growth. This process is generally undertaken by a government body that decides the skill sets required and the rules and regulations required. Communication, the web and emails are becoming a high priority to focus on the skill sets of a business and how a business should move. Google and working papers are becoming an important source of motivation. It is also very important to develop economic and legal framework and see to it that there is no deadly conflict. Very often during the process of forming goals, organizational myopia is created which is found because of tyranny of the inbox, media pressure and methodological shortcomings and the barriers for achieving these goals are political motivation, lack of leadership and vision. The ways of bringing long term goals to a success are strategies that can keep options open over time, decrease uncertainty, enhance continuous learning and allow better control and use of market and other information signals. It is found that single goal setting is easy and there are no problems on behalf of the decision maker while multiple goals setting can affect job related tension and performance. There are a lot of human constraints in managing multiple goals setting and help in the perception of organizational goals. It is important to prioritize goals and make trade-offs between them, where the manager does not suffer any anxiety that might result from such decisions and the uncertainty derived from such decisions. Uncertainty is a result of setting far sighted goals which can be achieved through allocating resources between the goal paths, actively working towards achieving goals and analyzing goal achievement. Uncertainty through multiple goals reduces performance and creates inappropriate decisions and it is a mangers' duty to understand the conflict between the goals. Productivity and job related tensions are hurdles whereas new technology and developing its prototypes are considered as long term goals. Stretch goals or high success goals are another management tool. Here it is necessary to identify or describe organizational factors or dimensions that have the potential to influence the implementation of stretch goals. Two of its dimensions are level of workflow integration and risk deviation. Stretch goals have autonomous work units and lead to considering a few sub-units for an integrated workflow. Stretch goals are formed by using linkage that helps in bringing relationship between the subunits and help in understanding workflow integration. They look towards output barrier, input barrier and variation barrier that through overtime, new hires and outsourcing retard the stretch goals. Gratitude oriented relationship, prestige oriented relationship, security oriented relationship consisting of protection against mistakes of business partners, protection against opportunism and cooperation oriented techniques consisting of flexibility, objectivity, freedom of decision and knowledge of customer are ideas of knowledge management. This would help in combining knowledge management to carefully calculated organizational goals. Support orientation is a very important phenomenon to manage complexity of modern businesses consisting of carefully manipulating the demand techniques like demand for advisory, demand for communication and demand for intermediation. Generally it is found that to describe in detail about the life-cycle of relationship based ideas that determine how long it would take to get to the next goal and when and to calculate how long the goal would last. You can think of acquisitions, intensification, reactivation through specific investment and recovery of the goal or business. Thus a goal ends with lack of CRM strategy or loss of customer. Both phases are different as CRM is getting very much technology oriented by building customer relationships and trust. But it can be found that the crisis of a product depends on loss of goals and vision. Thus goals rise because of introduction of successful business plan and later due to setting up multiple goals by looking at the time span and segmenting the product use and market. Design of relationship and processes through graphs and virtual drawing is the only way of understanding the future on a look at the present. Model creation very often creates business goals through the motivation of business process participants. They should understand compatibility and consistencies to fulfill all goals, including the creation of maps that link to each other very confidently without any misconception of the external environment. The external environment notation helps us to design alternative goals like design alternative, encompassing business process alternative and understanding of transformations, concerning organizations power and controls and operating quality. A business process is well defined only when there is inputs, standards and procedures for performing the work and understand the completion criteria. Moreover suggesting a more common goal that has high business autonomy should avoid cross process exchanges of information and material and understand operational costs in automated activities. Any business process under a particular situation may have a short cycle time where performance measurement is inside the business process and near to the performer. There are a high proportion of control activities, automated activities and human factors that affect the performance. After fulfilling the idea of developing on business processes it is important to look at the business benefit where it can be found that strategic marketing planning is the alignment of marketing objectives with organizational goals. It is development of a broad, integrated set of decision that ultimately creates the environment in which the firm operates and thus creates the major focus of a business. Corporate goals have been created through the idea of core competency where the research into the development of new technologies, have been extended to include technological consultancy and the organizations' wish.

Business methods have helped us to understand profit generation and profitability. It's marketing and sales can be strengthened in recognition of new challenges. A business process helps create scope that understands the competitive environment of the business, operate intuitively and dynamically, understand external environments affecting the business and look towards actionable intelligence.

As we think of goals in the normal phenomenon. The most prominent example of goals is of the project completion in a construction business. Similarly this project completion is the job of an expert who understands the differences between strategic intent and internal and operative management process without focusing too much on building related solutions. Thus this is a client's link between business development and facility planning. Innovation of collaborative forms is in focus as well as what can be learnt from, for example lean production. These ideas can be used as corporate strategies into property investments and corporate real estate decisions. The idea as to how Brazil became successful after overcoming the social and environmental challenges, creates an important idea of how corporate responsibility and related governance initiatives in the local and global levels addresses the social problems. There are many types of different social and environmental models where we think of companies as operating in a structure of multiple dimensions and interests. Here a broader set of stakeholders emerge which have a definitive impact on companies. From this perspective of business responsibility and sustainability, the responsibility of companies should be primarily most effective when it comes to the inherent characteristics of core business and related strategies looking effectively at resources, targets and rewards and from which major business impacts can result. It can be understood that environmental issues are more directly associated to tangible aspects of the business in terms of operational costs and related to monitoring and communication. Social issues in rationale not having a management rationale and are more significant transformations when it comes to consumption and production patterns.

Again looking at the major goal settings, consumer behavior like the pre-purchase goal rule out any uncertainty. This helps in processing more information and judge the value. Price promotions similarly have several benefits such as increasing demand, adjusting to fluctuations in supply and demand and increasing consumers' purchases over time. Price demands, marketers segmentation outlook and price discounts provide an important advantage to research on consumer behavior. This is a very brief factor of the usage of goals. To look more elaborately we also look at performance and business expansion as other key goal development areas for redefining a business where there should be control on speed of making decisions, focus on strategic and operational issues, develop new business opportunities and plan for impact on the organization and cost cut in whichever product that is beneficial. Far sighted goals of a business are through many steps where the time span may be up to 1 year and the main idea behind the focus is decided after the designing plans are made. After that, the focus is made on designing the new company for major challenges in the future and makes an analysis of Porters 5 force strategy to make it happen. To keep in mind the pricing ideas companies should keep in mind about the focus on narrow, specific strategy, and the heart of which is their specific market. This is linked to business strategy in terms of customers and markets and is a competitive dimension related to cost, quality, dependability and flexibility. There should be a formalized manufacturing strategy, relative to a non-formalized strategy where the impact of formalization can be understood. In this type of strategy it is important to understand the delivery speed and introduction of new products. All these are known as manufacturing improvement goals. Here the key ideas to deal with are quality related competitive goals and manufacturing improvement goals. These are related to cost where the un-discussed ideas include procurement cost, labor productivity, capacity utilization and overhead cost.

Goals are the competence levels which learners aim to attain. An intellectual quest has no terminal point in mind but linked to deliberate learning with a specific goal. The workforce are one of the key areas of competitive advantage where customer orientation, efficiency and quality help in understanding paths that lead to restructuring, reengineering, redundancy and revitalization. It becomes essential for organizations to align these organizational generic competencies to the behavior and performance of employees. As foresight helps in understanding that the competitive advantage of a business is superior customer service helps employees focus on the benefits of this strategy. To look towards the learning expertise it can be said that each generation of new products must be brought to the market quicker. New contracts won in the service sector must be brought on stream in an ever quicker timescale. Reactions must be swift to competitors in a globalised marketplace.

Goals are dominated by uncertainties which are also faced by uncertainty measured roughly by predicting the final outcome in terms of dimensions of time, cost and technical performance. The learning curve concept comes into use over here. Projects that are very similar to previous ones and about which there is abundant knowledge have lower uncertainty. Website addresses create a very important idea of goal setting where website addresses appear everywhere on printed and broadcasted advertising as companies scramble to inform customers that they have become part of the e-business phenomenon. It is very important to enhance the website and to integrate electronic business technology into the company's mainstream and supply-chain operation. Implementation of anything new in product, system or service also works on a project-by-project basis. The larger, riskier, more complex, costly, innovative, or different the thing being developed or implemented. Goals can be set in a number of ways by following the particular tasks. R&D needed to develop a product prototype and prepare specifications, applications engineering needed to define where and in what ways the product would be used. Marketing is required to define the commercial market and determine how to position the product. A new process has to be developed that would create a product difficult to copy. Finance is needed to determine the initial product costing, perform profit/loss forecasts and create policies for large scale production and marketing of the product and at last perform patent research. In the life cycle of a project that includes the phases of conception, design and development, fabrication and testing, installation and launch and finally enhancement, replacement or cancellation. Goals can be shaped in a number of ways right from the integration of a number of distinct firms in pursuit of a common investment strategy. It has been found that enterprises competing in major manufacturing industries, creates incentives for these companies to incur the high fixed cost necessary to attain competitive advantage.

CHAPTER 2

Gathering visionary leadership ideas and focus for sustainability

(BY BAISHAM CHATTERJEE)

In modern world and to lead the modern economy we think of visionary leadership. Vision helps to create environmental conditions of dynamism and hostility. Wider political, legal and societal issues increasingly affect today's organization. The problem of managerial obsolescence intensifies in scope and scale. There is need for reorganization towards improved environmental responsiveness, incorporating greater consumer satisfaction and more efficient operations-the hallmarks of quality management. It is important to formulate leadership dynamics and re-conceptualize leadership paradigms. Vision helps the strategists to concertedly explore the quality domain with a view to contain the complexities of modern business operating conditions and providing a counterbalance to the increasing vulnerability of market position. There is always a linkage between markets, competences and capabilities where the key characteristics of the person who drives vision to success are imaginative, experienced, intuitive and analytical. The customer responsive, quality based operations of visionary led organizations can lock in customers, at least in the early stages of the new operation and create huge price premium based profit outcomes. Some potentially successful visionary leaders have failed to create

8

successful organization because they have emphasized a marketing led approach to the exclusion of the marketing controlled approach. It is important to understand performance mainly preventive quality control, because of its propensity to reduce costs associated with quality appraisal and quality failure that helps in creating massive savings.

It can be said that visionary leadership has impact on positive organizational outcomes, where many scholars have turned their attention on the interpersonal skills and competencies that are necessary for demonstrating visionary leadership behaviors. It is important to understand emotional communication skills through knowledge, outlook and expertise that create leadership performance and effectiveness. Emotional competencies, self-awareness, emotional expressivity and self monitoring are the leadership outcomes. Visionary leadership defines where the firm should reach and how to initiate self driven and self generated goals to reach over there. To reach individualization and the true perfection of a visionary leader it is necessary to generate emotional intelligence and emotional competencies. A 360 degree assessment and understanding managerial performance links the knowledge and feedback to better create a transformational leader. Stable environments provide few opportunities for change but a visionary and charismatic leader can measure environmental uncertainty, crisis and create an organic organizational structure. The new model of leadership is dynamic leader built over the visionary model, suggesting when various facets of leadership are more important. The visionary theory of leadership looks at task and people dimensions and can more be understood through self assessment models rather than statistical forecasting. There are many macro, interpersonal and personal level ideas that through a two way communication and people orientation can promote team-working and culture of excellence. The visionary model helps in understanding compelling vision and organizational change. A visionary leadership measure helps in measuring the behaviors, provide leadership that is effective, innovative and gets results and understand emotional stability and openness to experience. A visionary leader has influence and thrust over change because generally a visionary leader can determine the nature and rate of change and is expected to contribute personally by using professional, specialist or technical knowledge and skills to the sole problems and make decisions. A visionary leader has influence over others and it is high when a leader has a lot of staff and requires their support to lead the change and has personal responsibility for developing staff and ensuring that they are equipped for change. With few employees influence is less and the leader has very minute responsibility for the personal development of the staff. Any very successful business has history of innovation and leadership. Industry visionary entrepreneurial leadership theory incorporates a combination of visionary focus on product design and development, with the courage to recognize a firm's resource limitations and a willingness to risk financial ruin to achieve innovative performance and production goals. Visionary leadership is style and strategy coupled together with direction and planning with ability to create a vision that others can believe in and adopt as their own. Such vision is long term in its orientation (while market imperatives are short term). The leader uses vision to build a bridge from the present to the future of the organization having the capacity to communicate that vision for instance through the process of management by wandering around or MBWA and translate it into practicalities. The most effective competitors in the twenty-first century will be the organizations that learn how to use shared values to harness the emotional energy of employees. As speed, quality, and productivity become more important, corporations need people who can instinctively act the right way, without instructions, and who feel inspired to share their best ideas with their employers. That calls for emotional commitment. On the ability of enterprise leadership to establish an environment that discourages "opportunistic" short-term behavior. Opportunistic behavior by individuals is likely by definition to be counter-productive to long-term value generation (especially where the individual has been able to appropriate some of the value they have generated within). Lean manufacturing systems require a high level of trust because, for example the fragility of the system, which can easily be disrupted, calls for responsible behavior throughout the network upon which it is based, at all times. This applies equally to suppliers and employees. Moreover people are trusted to deal with problems, where and when they happen, at source. Lean manufacturing is successful through this visionary leadership because: people are trusted with high levels of responsibility and discretion at all points of the supply and operational process. This implies a significant degree of the delegation of authority and responsibility throughout the workforce and the supply chain. The use of collective and team/cell-based operational structures means that free riding behavior becomes unacceptable. Group norms become dominant particularly if pay is also based on them over individualistic priorities. Moreover the abandonment by employees of traditional lines of demarcation and trade union involvement in the establishment of work practices must be reciprocated by management. This may mean making available the necessary multi-skill and quality assurance training; providing employment guarantees (at least to core workers); implementing single employee status, and the downgrading/ elimination of hierarchical privilege; increased remuneration resulting from increased productivity, etc. Moreover, there will be an expectation of totally cooperative and trustworthy behavior by suppliers and intermediaries throughout the value chain. This is related to the requirement that open information flows are needed to make the system work. The free exchange of information will only occur where there is adequate trust between the parties to that exchange. This is particularly true of the supply chain; and of network structures/ relationship architectures.

Visionary companies are characterized by strong drives for exploration and discovery, for creativity and innovation, for improvement, and for change. Visionary companies tend to be more demanding of their employees and managers than other companies. But those who can cope may develop a strong sense of working for an élite organization, which in turn has an effect on the caliber of people who can be attracted and recruited. Visionary companies may be regarded as ultimate employers who are in a position to recruit "the best". Visionary companies exhibit high levels of action and experimentation—often unplanned or undirected—that produce new or unexpected paths of progress. This evolutionary progress is opportunistic in character; accepts the value of trial and error and chance discovery and rejects 'Not Invented Here' limitations on the strategic management of technology and innovation. Individual employees are encouraged and empowered to seek new paths and new ways of doing things.

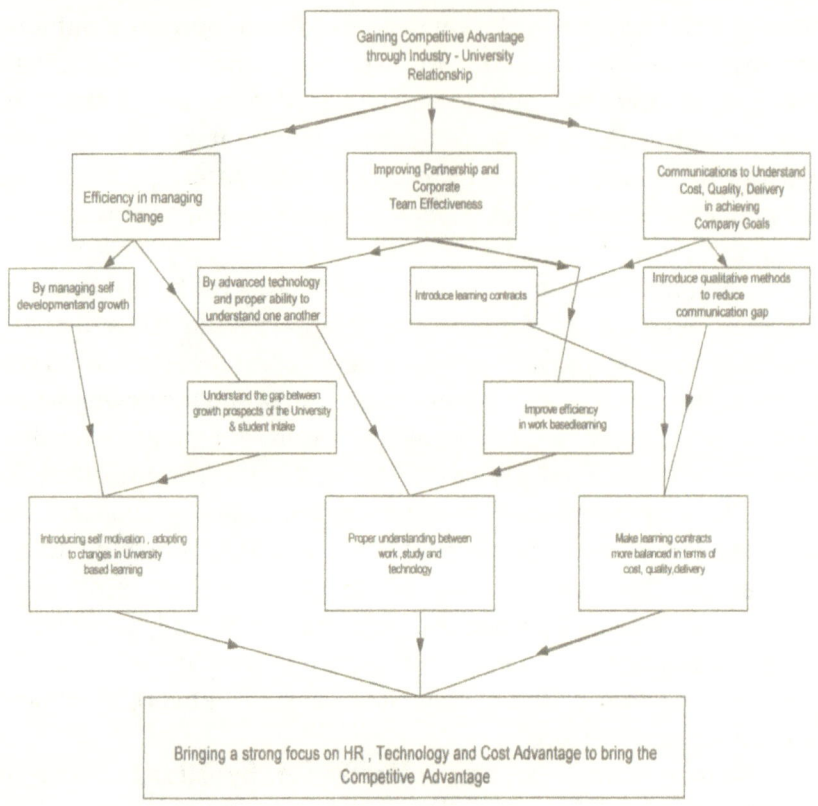

Fig. 2.1: Industry – University Partnerships (**B. Chatterjee , 2011**)

In the above fig 2.1 descriptions have been made on industry-university partnerships which gather competitive advantage through technology and cost advantage to bring the major focus in collaboration and vision Visionary goals can be created through bringing together IT support and organization where we can devise and implement a plan to maximize the competence in each role and separate the employment factors and steps into different roles and the leader and follower overemphasized the hard roles of technology and administration. Visionary leadership is built through a competence and is committed to the organization and to a purpose. Different theories help us understand visionary leadership through extraordinary levels of follower motivation, admiration, respect, trust, commitment, loyalty and performance with frame alignment, role modeling, image building and exceptional risk taking being the key ideas to derive success.

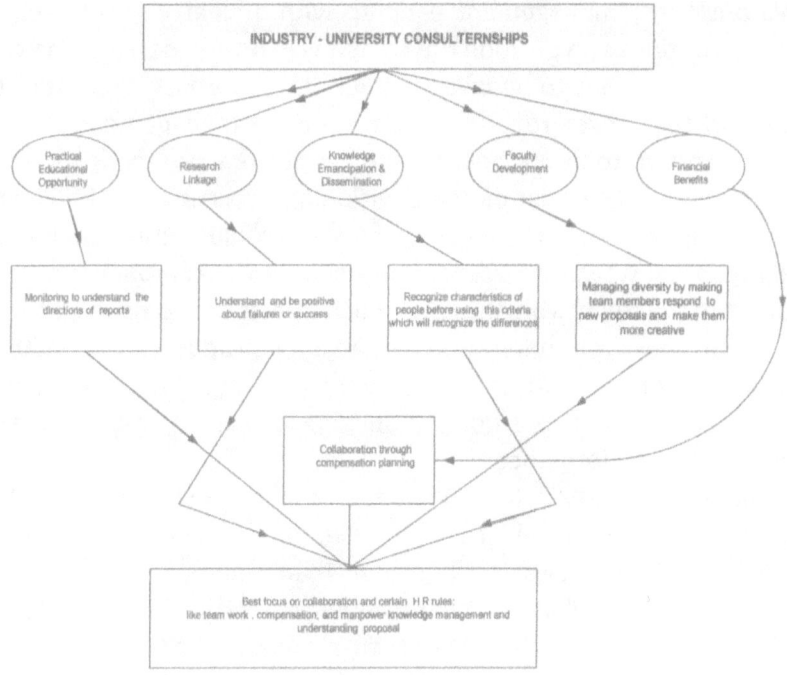

Fig. 2.2 : Industry – University consulternships (**B. Chatterjee** , 2011)

This fig 2.2 on industry-university consulternships, discussions have been made on how collaborations develop HR rules, manpower knowledge management, compensation and teamwork. It gathers different ideas that would develop relationship in the future. By modern outlook taking a visionary decision means having the highest potentiality within a leader. Decisions vary in three different ways: opportunistic, strategic and visionary. Where visionary decisions are applied during uncertainty and when environment is difficult to predict. The performances for envisioning a business are aimed at: maintaining excellence of performance and creating novel options. Visionary decisions are taken when there is requirement of excellence of performance, long term survival and there is requirement to find new options. The idea for bringing a successful visionary leadership idea is through creating new skills and creating new capabilities. Visionary management is always proactive. It strives to define new business areas for a company and comprises creation of new capabilities, organizational development and acquisition of new know-how and skills. Visionary management ends up with a clear vision to renew strategic options and position the company in future markets. The cost to the company has to be balanced with the improved opportunistic moves in the longer run. To realize visionary management, special communication tools are needed and various team-work approaches must be used. The goal of the envisioning process must be clearly defined and the work in the sessions must be conducted in a disciplined and organized way. In this sense the vision of the company is an end product of the team-work process, which directs the company into the future. Business reframing tools like Ansoff's product-market matrix, porters value chain and its five competitive forces, Ohmae's 3Cs and making an appraisal position by building a team membership and assessing the performance.

There are many rules and ideas behind service options that organizations can make. They are like professional development of employees to levels much higher than their expectations and aspirations, create jobs that offer psychological and intrinsic satisfaction and use competitive performance in such a way so that the standard of living of a number of people is elevated. Ideological vision transcends the self interests of followers; motive arousal of followers and the engagement of followers self definition; look towards self esteem and self worth towards the vision of a leader or the organization for which the leader stands. There is a lot of emotional strength in a visionary leader who has high energy and endurance; have a high desire to take risks and have good social adjustment; coupled with power, socially adroit and have self confidence. There are many constructs of visionary leadership like: The first transformational construct is intellectual stimulation, where leaders encourage followers to think creatively and approach situations in different ways. The second construct is Individualized consideration, where leaders consider each individual's needs and assist her or his development.

There also seems to be a strong visionary sense in managing non-profits because most of the non-profits are short of resources both funding and leadership talent. There are many characteristics and requirements of a visionary leader which are as when the milestones are achieved. Teams engage in discussion, exploring how they are going to compete in the global marketplace. They assess the strengths and weaknesses of themselves and their competitors, the stakes—what they have to lose and what they have to gain. Explore how they are going to shape the industry to their own advantage by exploiting their own strengths and their competitors' weaknesses by taking advantage of external trends and events. Similarly a business strategy is successful because customer research encompasses concentration mix, the potential for restructuring channels of distribution, new markets and/ or niche opportunities, relative buyer potential power, possibilities and consequences of forward and/or backward integration. Supplier considerations also focus on concentration, ease of switching, bargaining power of suppliers relative to others in the industry and forward/backward integration. Potential substitutes are explored with regard to how quickly they can penetrate, what impact they will have, and how other firms in the industry will consider them as opportunities for diversification. A potential leader can determine the target work that he has necessarily to do with stakeholders. It is found that many people who run government organizations, non-profit organizations and people working for administrative driven corporate tends to be idealistic visionaries with a strong focus of where his ideas would lead to in the future. This can be defined in two scenarios have sufficient capability and gather adequate resources to be idealistic, become tough pragmatists that would drive idealism. There are many organizational obstacles that require a strong sense of visionary leadership to tackle. The organizational difficulties are as: work load, interactions, leader characteristics, subordinate emotions, work environment, resources, and subordinate ability. With specific responsibilities within a defined region it is responsible for conducting needs analyses and devising and delivering custom-made programs to meet the flexible requirements of independent regions. Re-designing jobs and processes and reducing communication barriers are ways by which creative and very future oriented businesses are created through the influence of the most recent technological and strategy goals that can be well equipped from strategy research papers and books, as well as make a clear result of the decision taken by the government for businesses over there in the past and future. This shows a clear result of the certainty and environment and economic outcomes. Similarly thinking of the work load a lot of research on group work has to be made and ideas have to be carried out when it is successful and when it is not. Emotions and ability create the support and the major activity behind success in increasing performance and bringing in new talent and creativity. These two qualities help a business to remain sustainable in the long run and keep on providing results through its distribution of many short term goals that would dissolve the organizational difficulties. Moreover further difficulties which arise as threat can be dissolved through a sound knowledge of competitive strategy. The idea of visionary leadership that was innovative five years ago like decentralization, matrix management, etc is becoming standard practice. New twists, such as "adhocracies," continue to harness the creativity and energies of the work force. Other formats may continue to evolve. Moving operational responsibility, resource management, and innovation down the ranks has become the trademark of visionary leadership.

In 1980 Michael Porters viewpoint on business strategy and industry structure provided the foundational idea on how to create visionary leadership decisions by looking at the business strategies, performance and organizational culture philosophies which makes a clear implementation and environmental analysis. Although each area has been respectively noted as a major determinant of success and as being a significant indicator of performance, it is assumed that organizational performance is a subset of overall organizational strategic effectiveness. The systemic alignment of the organizational vision with the strategic vision requires defining the measurement for each respective core element. The identification of these measures will in turn lead to the development of systemic measurements that reflect the effectiveness, degree, or classification of success of vision driven organizations. Organizational researchers have developed a wide variety of models of performance measurement which include organizational structures, systems, and human resources but due to the diversity of conjectures and lack of empirically tested models, aggregation of the concepts is difficult. The advocated approach lacks the identification of a quantitative model that will predict or classify the degree of success in achieving the organizational vision. There are foundations to influence major decision-making processes concerning short—and long-term organizational direction. The researchers identify that vision driven organizations utilizing a systemic approach reflect a predictive pattern of financial measures. This predictive pattern can be utilized to classify the degree of success on the continuum towards becoming a truly visionary organization. Shifting the traditional strategic paradigms from the utilization of financial indicators as reactive measurements to those of financial indicators as proactive or predictors of success for vision driven organizational effectiveness, will be a major paradigm shift. The challenge for leadership is to accept this new paradigm as a long-term organizational developmental tool that utilizes vision classification for positioning on the journey to becoming a high-performance visionary organization. An effective vision driven strategy translates directly into organizational effectiveness which can be measured by sustained competitive advantage, improved sales, solid leadership, greater employee commitment, and increased shareholder value.

There are said to be three results on visionary leadership where the three aspects are intentional influence on behavior, building on effective organization and achieving efficient performance. A relevant environment reduces uncertainty and helps decide on action. This relevant environment should have the correct and well calculated production systems and should have more ability to do research on processes. A visionary leader always brings together people with a common cause. Vision is related to culture that creates a purposeful group that ultimately gives the intention to look forward. An individual or team helps in understanding a job or task and develops the initiative to create the most efficient performance. It is said that there are four types of leadership ideas that rule the environment: like strategic leadership, inspirational leadership, administration and supervisory leadership. Moreover training helps a visionary in transforming an organization for the good of people working in the organization or some other organization. Visionary aspirations apart from transformation or ideas that bring in transformation may be more directly relevant for middle managers, as their task profile includes translating such visions into specific operational plans. Visionary leadership from upper management may, therefore, directly impact middle managers in their daily jobs, providing them with a clear sense of direction and illustrating the significance of their efforts and the purpose of their work. Thus, middle managers' job satisfaction is likely to substantially increase through their leaders' inspirational motivation. In contrast, visionary aspirations may be less directly relevant for first-line supervisors. As opposed to the long-term perspective organizational visions entail, first-line supervisors' tasks usually have a short-term orientation, focusing on daily operating problems and on the management of individual performance. Organizational visions that focus on overarching goals may thus have limited relevance for first-line supervisors' day-to-day work and may appear abstract, unrealistic, and unconnected to their actual tasks. First-line supervisors, therefore, may derive only a limited sense of meaningfulness from middle managers' inspirational motivation.

It has been found that there is often a linkage between visionary leadership and its core values. There core values are the qualities of an organization, the qualities and sudden changes made in the value chain and the changes in the educational environment and learning that makes a leader. Everything rests on the core value propositions that can not only include the business alignment and qualities for sustainability, but also includes the ability to interrelate and form an internal organizational flow. These deeply ingrained values can be understood through transformational leadership that identifies the correlations and models and diagrams governing those correlations. Many leaders use leadership behavior in order to set standards, communicate their vision and mission to ideas and encourage their workers to grow and develop beyond the norm. Self determination is a process to create high value efficiency that helps in transforming their followers through communication, role modeling and encouragement which are appropriate strategies for achieving missions and goals. The tendency to control and plan is limited by its short-term approach that frustrates workers. Leaders who intellectually stimulate to build character as well as internal behavior skills by understanding the challenges and coaching. Technological breakthroughs, globalization and joint venture helps in understanding change build great achievement results and thus easily manage short term results. Understanding systems and processes is a critical idea of doing management. To understand the specific impact of leadership behavior the assessment of behaviors and their resultant outcomes found no consistent patterns across situations of relationships between leader behavior and outcomes. However, there were some consistencies within certain types of situations which suggested that the situation or context of the leadership process might impact a leader's effectiveness. The leadership factors that can be focused by a leader are business and organizational; thinking and analytical skills; innovation, tolerance of change, risks; team playing; leadership; customer focus; interactive and interpersonal skills; and personal characteristics. We very often talk of core prospects in visionary leadership. Challenging the assumptions of the core comes naturally to a revolutionary who lives on the edge of an organization, not at its center. Once embedded in the core, inspiration gives way to issues of continuance. The organizational core is an illusionary place of stability that takes effort, not inspiration, to maintain. Individuals in leadership positions become uninspiring as they spend more time in the core. These leaders continue to retain their positions by communication strategies they employ intended to ensure that they stay in positions of leadership. Many of the leadership efforts are not to lead but to maintain. The efforts of once inspirational revolutionaries are directed to maintenance of control rather than in inspiring the workforce. Uninspired leaders keep specialized management consultants in business by ensuring a stable set of working conditions in which they are always asked to come in and make sure that all around the uninspired leader are told of the visionary qualities of the leader and are asked to talk about the wonderful successes that have occurred or might occur in the near future. Morale stabilization is an important consideration in the hiring of this special class of consultants.

The quality that helps in motivation is generally related to the meaning of the business as well as the competence that is the first requirement for a visionary leader to understand. Self-determination lies beneath the work behaviors that are completely related to the experience, support and confidence of the visionary leader. Moreover impact of individual behaviors and its work outcomes determines the ability or the required input, effort and required perception of the visionary. This step helps in earning perceived control and perceived competence that helps in understanding the motivational needs and situational demands and the type and quantity of resources that are to be utilized for the particular possibility. Goal internalization that is an energizing element of a goal is a valued cause or meaningful project. If a traditional practice that makes an employee powerless is removed, it is believed that employees perform at their creative best and thus this might lead to empowerment. Leaders with vision create a much more influential climate with a more empowered condition than organization members. Besides providing a vision transformational leaders engage in inspirational behaviors that build subordinates self-confidence with respect to goal attainment. Management by exception behaviors for example focus primarily on mistakes or slippage of performance that generally have less symptoms in a visionary leader. But this form is existent in other forms of leadership. This inadvertently communicate to subordinates that poor performance is anticipated but they are not expected to take initiative to correct it. This helps in solving problems with different goal sets to overcome difficult objectives and look forward towards managing higher level of capabilities. In the case of a visionary leader and the decisions that he takes potential teams that the leader has recruited were more productive than those with less potency. Potent teams provide higher levels of internal and external customer service, with determination, empowerment at the individual level and define the experience by which team members have freedom and discretion in their work. Team members have externalized goals and objectives, related to the purpose and inspiration experienced by members who are committed to achieving work objectives, related to the sense of inspiration experienced by members who are committed to achieving work objectives which are clearly linked to valued organizational objectives. Goal internalization corresponds most closely to the empowering effect of transformational leader. A visionary leader can take a step ahead and create positive response by understanding the work team effectiveness that helps in understanding productivity, quality, low costs, safety, job satisfaction and organizational commitment. A transformational or visionary leader can create an organizational environment where followers feel empowered to seek an innovative approach to perform their job without a fear of getting penalized. By looking towards the impact analysis, it can be described that. Suppliers want to be known as the organization that provides a component to the new organization's offering. Partners want to be aligned with the new entity to appear fresh and vital. Stakeholders want to invest, in one form or another, in the new venture. Most importantly, existing and potential associates want to work for the organization, seeing it as the best solution for achieving their own personal goals. It's always fascinating to watch leaders who tend to move through this stage with unbridled passion and enthusiasm, but fail to recognize when organizationally impacting changes are occurring. Besides talking of vision we can bring vision and creative tension together. Creative tension comes from seeing clearly where we want to be, our "vision," and telling the truth about where we are, our "current reality." The gap between the two generates a natural tension. Without vision there is no creative tension. The natural energy for changing reality comes from holding a picture of what might be that is more important to people than what is. With creative tension, the energy for change comes from the vision, from what we want to create, juxtaposed with current reality. With creative tension, the motivation is intrinsic. This distinction mirrors the distinction between adaptive and generative learning. However, vision is more than an image of the future. It has the power to inspire, motivate, and engage people. Vision rallies people for a joint effort, motivates them to become involved and committed, promoting quality performance, causing them to exert additional efforts and devote time to devote to organizational learning processes, aimed at improving school outcomes. The present model and findings also offer a rich agenda for practice. The educational system as a loosely linked organization faces difficulties in creating coherent activities related to school performers. Therefore a study model, indicating a relationship between school leadership, vision and school organizational learning, provides coherence and co-ordination, enabling the necessary interaction between the staff member.

MODEL OF SUPPLY EXCEEDING DEMAND

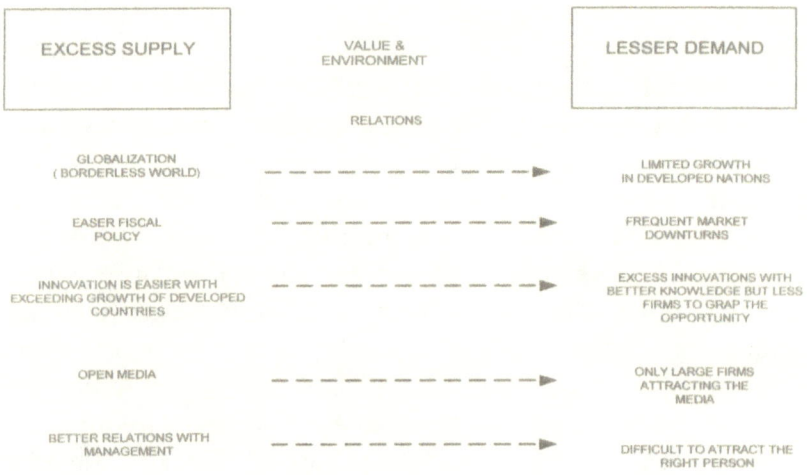

Fig. 2.3: Model of Supply exceeding demand (B. **Chatterjee**, 2011)

In fig 2.3 on model of supply exceeding demand relationships have been developed on excess supply and lesser demand which suggests markets and economies that have lesser demand of a particular product and economies that manufacture and export products more than its requirement. Thus this figure states the elements of each. Innovation, vision and goals are interrelated. Innovation has become the key goal of many organizations because of its potentially significant impact on organizational performance. With the ongoing social and technological changes, it is argued that innovative efforts of individuals at the workplace have important impacts on organizational performance. They further posited that leaders who define group goals and control critical resources are able to create contexts and conditions that motivate followers to engage in innovative efforts to accomplish their goals. Studies found that it takes leadership to create a climate that is conducive to employees' innovativeness and creativity. Leaders have significant impact on how followers go about achieving goals. Research found that transformational leaders who align the values of followers to their own and to the organization's are able to increase their followers' intrinsic motivation much more than other leadership styles.

Leaders have influence in developing risk based strategies that help in performing and thus bring consistency in behavior. It is charisma that makes a leader visionary, motivational and powerful, confident and captivate people. Charismatic leadership derives clear vision through needs of followers. Leaders display stimulation when they help their followers develop new ideas, motivate them and sole any potential and internally or environment oriented problems. E-mail and video conferencing helps visionary leaders understand the nature of group behavior and understand the social interactions required. Similarly project completion mainly in very large scale businesses requires a lot of visionary leadership ideas. Thus vision and visionary leadership play a prominent role through innovation, goal achievement and understanding the forecasting techniques.

CHAPTER 3

Incentives: future vision and theories
(BY BAISHAM CHATTERJEE)

It has been found through McGregor's approach through theory X and Herzberg's KITA approach to motivation about the compensation systems should be directly related to the company's sales prominence. Monetary incentives directly affect goal setting and if rewards are contingent on goal achievement, then lower goals lead to lower performance but certainly income helps in setting original goals that leads to bringing a prominent incentive structure. Incentives are somewhat a part of management by objective perspective. It helps in quantifying performance, performance linked to appraisal and compensation systems and understanding rewards. Incentives are said to be an integral part of strategy and performance and communications.

It is said that salespeople too deal with forecasting ideas where salespeople are informed by superior of preliminary forecast adjustment factors and the salesperson revises the forecast. This is a great area to develop upon where the salesperson is made well aware of the time management decisions and the salesperson wants to earn a higher rating related to forecast and achievement. Other valuations for incentives are de-capitalizing readily quantifiable lump sum type incentives, such as cash payments or free fit-outs into effective annual rentals. It is also very important to note that incentives which are dependant on future market forces affecting forecasted rental levels estimate the benefit in future. Marketers have been using WOM incentive programs with a variety of tasks or goals for the WOM provider. Some programs offer incentives only if new customers actually sign up for a service, others provide incentives for simply forwarding e-mail messages. WOM provider needs to be motivated to exert the required idea of the incentive offered, size of the incentive and consumers likelihood. Excluding the WOM that comes from within another impetus in incentives is transformation of production in the implementation of different technologies that would solve environmental problems. To undertake this idea there needs to be technology changes which relates to changes in organizational context and in context related to the technology.

Before going to the idea of linkage it should be understood that firms which clearly separate the process of decision management to understand various levels of management, regular understanding of the implementation policy would be necessary. This would reduce the scope for discretionary behavior. Moreover managing the divisional manager with a remuneration package helps performing better.

STRATEGY EFFECTIVENESS AND FINANCIAL INCENTIVES

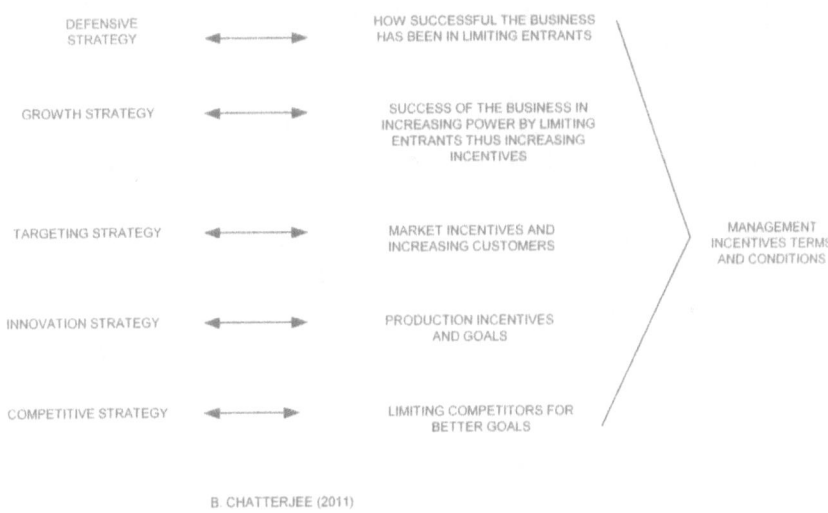

B. CHATTERJEE (2011)

The other very relevant idea behind incentives is e-governance and e-democracy. A new research focuses on e-democracy, in relation to accountability and its role. New technologies can play in the disclosure of government financial information. In most developing countries no good match exists between incentives held out by governments to promote development and perception of their own needs held by poor masses who ought to be the main beneficiaries of development. Web incentives are becoming an increased priority in the modern world. E-governance has five main aspects: interaction among government agencies, web-based service delivery, e-commerce, digital democracy for more transparent accountability of government and e-finance. E-governance helps in understanding the different ideas behind bringing changes in political competition, cost of debt, fiscal pressures and thus makes it easier to understand the steps and dimensions the community under the government should take. Other than this there are more important parts of incentives that are used to incentives for secret-sharing in self-organized networks that provides functional and technical details on the proposed architecture and through telecommunications. It assesses its real time experiments. The ISSON or incentives for secret sharing in self organized networks is applicable to dynamic environments. It combines positive and negative tokens as incentives, secret sharing techniques and pseudonyms to increase the availability of critical information to avoid the exposure to confidential information and to enforce unlink ability. Using ISSON peers could choose the secret way they want to share, the level of privacy that is associated with they want to share, the level of privacy that is associated with the sharing procedure and the entities that will be delegated to maintain and reveal the secret. ISSON helps in secret-sharing and the authorization and information distribution process. ISSON introduces a credit/denial incentive mechanism that enforces peers to participate in the sharing of classified information. There is much information generated terms of ISSON that can lead to developing a peer client, peer STS server and client lists and understand client id and issued credits.

Joint gains are an important possibility of getting joint outcomes. Early research found that shared information increased the likelihood that bargainers would reach an agreement that maximized their joint outcomes. It helps in understanding the measures of bringing out developments in joint profits. Information seeking and negotiating is becoming an integral part of this joint agreement. Opportunistic behavior is another form of incentive where opportunity starts from negotiation. Negotiation structures can be from very simple to very complex issues when negotiators with conflicting preferences must negotiate a single issue. Opportunity rises from competitive measures and a mix of distribution and co-operation. Incentives may be quoted in various forms and moreover the incentives that are non-financial and the most important part of that is job enrichment. Job enrichment may be organized in various forms by factors like job re-design, job sharing, flextime, participation and team focus. Job enrichment can be built by various inquisitive ideas that come through years of personal experience in various hierarchies and at various types of jobs. Such people have a better idea of how the job can be designed and how it can be made more attractive and knowledge-sharing. After this phenomenon of job enrichment the idea of job sharing can be brought. Giving autonomy to employees is a very important idea practiced even in 3-4 member businesses as well as creative businesses which needs internal leadership and lot of personal and inter-personal nature to be successful. After job enrichment and idea sharing and building a superior team effort and understanding in the organization comes the study and advancement in understanding job advancement, empowerment. This helps in creating superior challenges for other businesses as well as helps in creating a destination and framework for their challenging employees to reach over there, taking it as a personal idea and self motivation skill. Treating employees with respect and worthy pay is also very important.

With time the recognition and rewards program should be restructured and reengineered that is a highly self-understanding and elaborate and costly system that may be carried out in such a way that it is administered in order to reward employee achievements that are indicative to commitment to the organization and are tied to specific accomplishments. There is more to talk of job enrichment in personal as in organizational behavior. However it is smaller firms that are unable to compete via compensation with larger firms. However it is the smaller firms' best interests to attempt to establish perceived internal equity. With regard to external equity which is not a predictor for job satisfaction, it is still relevant, but small firms need to communicate objective factors to understand the statistical idea that marks the remuneration marking based on performance.

Innovation is the lifeblood of a business and the business should find ways to enable the company to the pioneer and drive high margin markets and find out new ways of manufacturing and distribution. This raises serious questions about the link between teams and innovation. Although work-group processes may provide ideas for individual employees to consider, which they would not have known of, if they had not been a member of a group. It is often the individual mind that comes up with the ideas that the group may refine. The more employees feel that they are being involved in company-wide decision-making; the greater is their commitment to the organization. And when employee innovation could save the company millions or lead to a revenue-spinning new product, the argument for mobilizing the whole company to innovate could not be stronger. Political stability that would make a preference of the geographic location is another incentive. Other ideas to elaborate on are infrastructure quality that would help in understanding whether the site or location suits the resource mobilization and the area can create value addition. Creating products for a very large market size is an incentive to grow faster that also helps in adjusting the cost which is very well considered to be an incentive.

There are many different techniques on which investment incentives can be measured that consists of ideas that deal with labor cost, land cost, energy cost and corporate tax incentive. The second factor consists of ideas that deal with local capital and relevant firms support. The third factor consists of labor quality and skilled labor force. Factor four consists of customs procedures. Factor five consists of openness of foreign labor policy, local regulatory environment and guaranteed foreign investment policy. Factor six consists of political stability factor, security and efficiency of government administration. Factor seven consists of location and transport factor that governs transport linkage and geographic understanding. Cluster analysis stands as a key factor in incentive generation and has many parts under it which are as firms that preferred political stability and location factors, port related factors and agglomeration effect. Logistics helps administrators of logistics zones to analyze investment incentive preference segments and modify their current policy to more accurately meet investor's requirements for the availability of firms support and organizing a labor utilization dimension. Logistics zone administrators need to determine type of industry desired and then match the incentives with the needs of targeted industry. It is an important part of developing marketing efforts and incentives to investor needs. They should first improve government efficiency and then provide a low-cost environment.

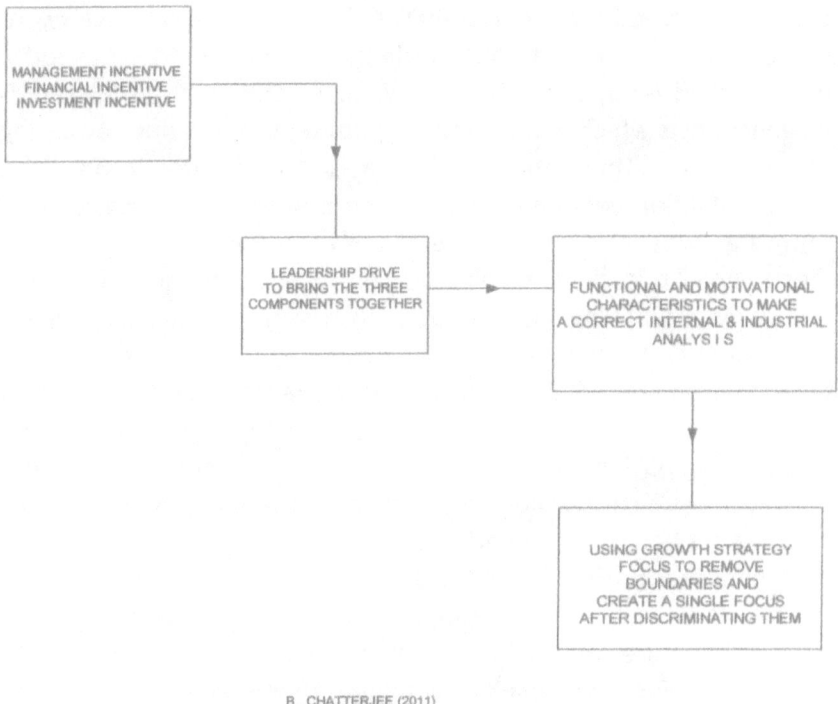

B . CHATTERJEE (2011)

Compensation programs are other means of incentive management taking into view the question of increased pay for increased size of business phenomenon, growth rate, the ability by which the other resources can be utilized to grow the business etc. Weighing the short term and long term performance measures of the business helps in understanding the compensation packages. Long term incentives are based on 1 year earning results for chairman and CEOs and profit center managers. Market based incentives comes from this basic concept of incentives, where it is important to understand the incentive portion on market returns realized by shareholders. Movements in company's stock price may well be influenced by factors beyond management control, such as the overall state of the economy and stock market. Second, market returns may be materially influenced by what management believes to be unduly optimistic or pessimistic expectations at the beginning or end of the period.

Performance based incentives should be offered to employees in order to gain maximum advantage regarding the growing prospect of technology and this type of incentive helps a person to move forward and compare with other businesses in the modern world. Firms today are characterized by separation of ownership and control which leads to agency costs which many firms have attempted to reduce by basing management compensation on firm performance. Net income and market valuation definitely affect a firms business, but it is such that a firm's growth rate, growth rate of the economy, financial rates all affects and all this can be regulated by proper location measures of big businesses and staying targeted to the growth rate of small businesses. Management's compensation can probably be graded in many ways depending on credentials, previous performance, qualifications that help in comparing with peers and understanding what is to be done in future. Big businesses can use compensation committees whereas small businesses can use personal salary benchmarking and a provincial authority for better benchmarking.

When the business becomes big management buyout is an objective that helps in preventing a non-core unit to succumb or prevents the core unit to function badly. It can help in increasing a firms own performance due to the managements years of experience and helps in better use of benchmarking to make things better. The first benchmark is against a matched firm of similar size in the same industry and the second is relative to the industry average performance of firms. After a buy-out the management team holds a much larger stake in the business and tries to create directions for further development. Financial incentives is what that creates the maximum motivation possibility of a person and there can be various types of financial incentives that can be grouped and organized in different segments and their credentials may be identified in ways that can put pressure on the management board to gather decisions and put a single focus on them to lessen distractions. The amount spent as incentives should be linked to the nature of a business and how effective the business's strategy has been.

Procurement is another area of deriving incentives. The prohibition of transfers from the government to the firm is not a relevant option under procurement. Similarly incentives and quality are an important area to deal with where, the demand for the good is a proxy for its level of high quality and incentives. The high quality can be provided by rewarding the firm more when sales are high. Incentive schemes are positively correlated with the firms cost performance and issues a warning concerning causal interpretations of this relationship. It shows the optimal cost-reimbursement rule and more generally any scheme that gives incentives to reduce costs. In planned economies and regulated industries there is more chances of having the best results from incentives. If the firm performs well early in the relationship, the regulator infers that the technological parameter is favorable and tries to use an industry-wide demand analysis to extract the firms rent. Thus firms are not indulged in bringing low cost actions, but they indulge in giving rewards for good performance.

A gift is given as a corporate gesture in recognition of business conducted or as part of business to business promotion. Incentives are products or schemes perceived to be of relatively high value and typically used to reward staff or long standing business customers. Companies that wish to purchase corporate gifts and incentives can choose from a seemingly endless range of items, which can be placed into three broad categories, based on the value of the item and purpose of the gift. The first category are give-away's, generally low value, high volume, less personal items that are used mainly to promote a company's name. Without the incentives, the salesperson would not exert pressure and play game with the customers. Some independent dealers have also eliminated the commission based pay. The salespeople will be paid with a decent base salary plus a bonus based on the number of items sold. Since the bonus is based on the number of items sold, instead of profit margin, the salesperson would less likely be playing the pricing game with customers. Other than the direct cash compensation method, dealerships should also look into other rewards. Training, opportunity for advancement and free use of demonstration models are some of the more popular non cash rewards. Many dealerships have started giving their salespeople profit sharing as recognition for their hard work. This will make the salespeople feel like a part of the dealership instead of just a tool to sell products. Of course, the best way to reduce the turnover rate is to hire the right people from the beginning and to treat them right. The dealerships should change their compensation method for the better. The current internet technology has allowed dealerships to reach customers beyond their geographical coverage area. They do not only compete with other dealers in their area but also with dealers from all over the nation. It is not uncommon to hear someone purchasing say a vehicle from a dealer from other states just because they offer a lower price. For these reasons dealerships have to treat their sales force well so that they will pass this favor to the customers, making car buying a pleasant experience for the customers. Incentives fall into three groups: material, social, and activity. From the simplicity of straight salaries to the complexities of stock option programs, compensation packages are a subset of the broader class of material incentives. Other such incentives include the larger and better furnished offices of senior executives or the additional perks received when joining the executive ranks. While these incentives convey a message and prestige, they do not necessarily draw the executive's attention to the appropriate corporate goals. Other material incentives that can provide more specific guidance include access to technology and information. For example, installing the long-awaited computer system generates enthusiasm on the part of the CFO, who can better project the revenue stream, or the vice president of operations, who can better track inventory. Social incentives, which operate on the interpersonal level, include identification with the company or recognition by peers, customers, and competitors. Social reinforces can be as effective for executives with a high need for control as those with a high need for membership. Activities that serve as incentives are those that are so rewarding that they provide the necessary reinforcement to sustain an executive through the more mundane tasks. For executives with a strong need for accomplishment, such incentives enable them to see their results e.g., bringing a product to market. For those with a strong need for personal growth, these activities include new challenges and opportunities. At the corporate level, the incentive strategy links various motivational needs with business goals. To satisfy the strategy, the incentives must meet two criteria: (1) they must reinforce each of the motivational needs; and (2) they must reward activities and behavior that support reaching the business goals. If, for example, the business strategy is built around providing quality goods and services, then the incentives need to reinforce this message. The vice president of marketing may have a strong membership need. In developing a marketing plan that promotes this strategy, he will rely heavily on staff meetings, internal committees, or informal conversations with his counterparts. Alternately, the vice president of operations may have a high control need. As a result, in running day-to-day operations, he will develop clear quality standards, regularly monitor adherence, and provide feedback. Two types of incentives are outcome based and compensation system based, where the first one may be costless or complete information. Therefore, normative agency theory would suggest that the board must either use outcome-based compensation plans or engage in a costly monitoring of management's decision making as a measure of behavior. In practice, polar cases rarely exist. Therefore, some information useful in monitoring a manager's behavior will generally be available. But since perfect monitoring is impractical or infeasible, it is also likely that compensation based totally on behavior will not be feasible. Therefore, some level of incentives also may be required if the principal is to balance the need for monitoring with the cost of information needed to do the monitoring. In fact, this scenario depicts the choice-continuum for incentive system design. At one extreme are systems in which outcome-based incentives dominate the compensation scheme. Systems at the other end of the continuum are based primarily on management's behavior. Compensation systems which are based on outcomes typically employ performance measures such as profits or return on investment which is readily available from corporate financial reporting systems. It suggests that part of the appeal of outcome-based incentive systems to principals is attributable to the ease of developing measurable performance data from information which is already captured in the company's financial reporting information system. Under this type of incentive system where managers are evaluated and rewarded solely on their ability to maximize reported financial outcomes, whether the decisions and actions were appropriate at the time they were made is given little weight in performance evaluation. Thus it is the reported financial results that govern the compensation plan. Further, compensation is tied to outcome measures through the use of objective, formula based criteria. Total compensation may be dominated, for example, by bonus payments made after results outcomes for the period are determined. Criteria may also be based on budget targets or business segment's results rather than overall corporate performance in an effort to tie compensation to financial performance which is controllable by the manager.

Incentives are regarded as a motivator nor incorporated into MBO sales forecasting systems.

This proposes that middle-term sales disciplines should be quantified and assessed by means of a merit-based appraisal scheme. Job characteristics are defined and then prioritized by means of a weighting system. Wherever possible, the model seeks to involve salespeople in a participative forecasting system which is intended to promote a sales culture. In this effort, effectiveness and accuracy obtain recognition via the annual salary award. Moreover incentives can be managed in health sciences and any other part of a business and compensation is a very vital part of a business.

CHAPTER 4

Best practices and balanced scorecard: an internal resource utilization process
(BY BAISHAM CHATTERJEE)

Internal transformations include using ideas like JIT, TQM and other processes. Similarly corporate performance is positively related to the role of manufacturing managers in strategic decisions. Implementation of best practices leads to superior performance and capability. When practices are examined individually, companies using any of seven practices (FMS, CE, Benchmarking, TQM, JIT, Manufacturing cells and Computer networking) have higher performance than those not using them. The classification of manufacturing practices taxonomies developed provides insight into the role of individual practices, implementation and outcomes. Time-based manufacturing practices tend to lead to standardization, formalization as well as integration. A structured approach used to identify direct qualitative relationship between practice and performance, integrating customer performance measures with internal performance measures (internal quality, productivity etc.) to identify improvement opportunities is found to be critical. There are only few best practices contributing to competitive manufacturing performance in multiple dimensions. It is important to update the company's process equipment to industry standard or better mainly by expanding manufacturing capacity and engaging in process automation programs. It is important to reorganize the company towards e-commerce and e-business configurations. It is important to rethink and restructure the company's supply strategy and the organization and management of the company's supplier's portfolio. It is important to assess process focus and streamlining of a business and understanding pull production and quality improvement and control.

The best practices of the most significant areas of study has to be understood like TQM and quality movement which consists of eliminating the disadvantage of quality control inspection, using natural and skilled work teams, empowering the workforce, bringing in quality and benchmarking into product and service. It is also important to calculate cycle time analysis, process value analysis, determine process simplification and undergo a strategic planning framework and formal supplier certification programs. Organizations that made frequent use of such practices depend upon the different process aspects for turning a business into a high performing organization. Organizations can also understand the strategy from the middle management to gain competitive advantage with positive impacts on profit, quality and productivity. There are many ways in which a business grows from low performing to medium performing and to high performing businesses. Cross-functional teams that use input from customers to develop design specifications for new products and services help in internal customer complaint systems. If the organization type is medium it is important to listen to supplier suggestions about new product/services and select suppliers through combination of certification and competitive bidding. To develop a business into a high performing business it is important to have world class benchmarking information, increase process simplification and cycle time analysis and benchmark marketing and service delivery. Communication between suppliers and customers understand and develop the communication criteria in after sales service. All this is a case of process development and analysis.

To understand the strategy and technology impact it is important to emphasize cost reduction when acquiring new technology. Using public domain as a source for process technology makes heavy use of customer satisfaction and strategic planning process. As the business turns into a medium performing business the firm has to make regular and consistent measurements of progress. Providing information to mid-management about the business consequences of quality performance and emphasizing the creation of more products in expansion plans emphasizes quality as key to reputation. In case of a high performing business an organization should look at competitor comparison, quality, reliability and expand geographically. Quality and team performance and selecting suppliers based on general reputation, maintaining design specification and shifting primary responsibilities are for compliance with quality standards. Downsizing the business and offering fewer services, focusing on cost reduction to make decisions about acquiring technologies are the main areas of a medium performing business. The process of benchmarking and best practices has passed from a continuous and systematic process of evaluation of the products, services to a continuous process of identification, learning and implementation of best practices in order to obtain competitive advantages, whether internal, external or generic. The other difficulty finally relates to the identification of these practices. And that just as much regards the entity side companies, production plant, service, etc., wishing to acquire these best practices. On one side in fact, the entity generally encounters difficulties in formulating its requirements about practices. The objective of the benchmarking which it undertakes is precise for it to discover new practices that are radically different from its usual practices and generally not yet considered. On the other side of the exchange, the entity generally has difficulties to identify, among its usual practices, those which produce significant profits in a context of application different from their own. In order to achieve the, core competencies for the future were identified; methods were then put in place to ensure that employees developed so that the company's vision of empowerment could be realized. As a result of this planning, the culture has significantly changed and enabled the company to remove a whole layer of supervision and management on the shop floor. To replace these roles, five shift teams have been introduced consisting of employees with a good mix of skills covering engineering, quality control and operators. This team-working initiative is based on encouraging total involvement, total responsibility and increased flexibility through no demarcation of the workforce. Although there is no supervision, facilitation and other support services can be accessed when required.

There are 2 types of benchmarking: Competitive benchmarking which refers to collecting specific information about competitors' products, services, processes, strategies, and business results, and making comparisons with those of the benchmarking firm. Co-operative benchmarking that focuses on sharing experiences with, and identifying best practices of organizations, which are willing to co-operate. These organizations may, or may not be, the direct competitors of the benchmarking firm. The rationale of benchmarking is to research and identify better best practices and/or performances.

This has proved success levels in a specific area, and which have the potential for improving other similar current practices and or performances. In general, benchmarking enables and/or motivates one to (a) determine how well one's current practices compare to others' practices; (b) experience best practices in action; (c) locate performance gaps; (d) prioritize opportunities and areas for improvement; and (e) improve current levels to world-class standards.

There are few best practice factors like codification, complexity, compatibility, perceived operational benefits and costs. These are directly linked to organizational factors like satisfaction with the existing practice, organizational resource availability and existence of champion which are again directly linked to environmental factors like perceived outside support and external pressures. Supplier searches and progress reporting is another form of best practices and are generally used in manufacturing firms and can be termed in a very elaborate way. The company consistently uses the Internet for effective and efficient supplier searches. This is usually highly successful, as there is a wealth of information to be found and it takes very little time to access it. However, one should screen the information found to avoid being inundated also by useless information. The products delivered by one of the benchmarking partners consists of more or less two sets of components: Components based on the core technology produced by the company, i.e. which gives the products their functionality to make them competitive and a surrounding framework of generic sub-systems like safety bars, platforms, etc. made out of steel. These latter components are rather generic and can be manufactured by a very high number of different suppliers around the world. Since these parts also represent either large volumes or high weight, they result in high transportation costs if manufactured by the company itself and then transported to the installation site. To lower these costs, the company searches for suppliers of such components near the customer's facilities. For parts bought from suppliers that are complex and require a rather lengthy manufacturing process, it can sometimes be very difficult to know at what stage in the process components are available. This makes it difficult to assess whether there is a chance for delays of parts being supplied from external suppliers and what the consequences might be regarding the company's delivery time to its customers. To overcome this uncertainty, a visit to one of the companies has to be implemented as a procedure. It requests weekly progress reports on such parts from their suppliers. The reports give a brief status overview that can be used to make schedule adjustments if progress is less than anticipated or even speed up operations if parts can be acquired earlier than first estimated. This gives flexibility in the production planning and eliminates some of the problems related to critical parts and scheduling of connected operations.

There are many different ways of building relationship between the customers through control relocation, contact relocation and integration. The main idea would be integrating business process with suppliers and reduce the number of contacts with customers and build multiple other interrelationships. The product and operation view would be regarding tasks relating to the same type of order that distinguishes new business processes and eliminates unnecessary tasks of a business. The behavioral task would be regarding relocating tasks to appropriate places. Consider whether tasks can be executed in parallel with decisions, design business processes for typical orders and look towards the external environment. This is a way of considering outsourcing as a business process and considering a standardized interface with customers and partners. There are different dimensions of manufacturing flexibility or best practices which can be determined by implementing manufacturing flexibility. Specifically identifying flexibility dimensions requires investigation, measuring gaps, selecting methods for closing gaps, and continuous assessment. In the first phase, identifying flexibility dimensions requiring investigation, senior managers must identify the specific aspects of flexibility e.g. range and speed they believe are necessary to compete. An example from the telecommunication industry, where the time aspect of product flexibility is required to successfully compete, as being the first to market a successful product is almost a guarantee of substantial market share. In addition to identifying the required flexibility types, managers must also determine if excess flexibility is desired. Excess flexibility may be desired to redefine uncertainty or to bank flexibility. In the second phase, measuring gaps, the required, potential and actual performance levels of the desired flexibility types are identified and measured.

Required levels of flexibility are derived from managers during the first phase and by surveying customers. The assessment of potential flexibility is determined by manufacturing engineers, while the actual levels of flexibility is obtained from the performance data of the organization. During the third phase, selecting methods for closing gaps, managers address any gaps between the required, potential, and actual level of manufacturing flexibility. Any gaps identified are prioritized with managers determining if the potential gains from achieving the desired level of flexibility are greater than the costs associated with obtaining it. The final phase, continuous assessment, ensures any required, potential, and actual flexibility gaps are being closed and that the required flexibility is still needed. This phase is achieved by the implementation of effective measurement systems and a top management steering committee with the primary function of detecting the need for changing which flexibilities are considered relevant.

Moreover it can be found that leadership can create performance driven best practice by bringing a strong customer focus in vision, mission and values and making business strategy be driven by customer and market knowledge. Generally to take the best example of best practices a supply chain maturity model can be developed. Attempts to advance without a base of firmly established practices are rarely successful. These established practices are:

Stage 1. Functional departments within an organization focus on improving their own process steps and utilization of resources. Managers typically focus on their individual departments' costs and functional performance. Managers who typically do not define well or understand well the processes that cut across multiple functions or divisions, resulting in limited effectiveness of complex supply chain processes.

Stage 2. Managers now define division—or company-wide processes, allowing individual functions to understand their roles in complex supply chain processes. Managers clearly define cross-functional performance measures and hold individual functions accountable for their contributions to overall operational performance. Resource requirements are typically balanced across the organization. A well-defined demand supply balancing process that combines forecasting and planning with sourcing and manufacturing is evident at this stage.

Stage 3. The company now extends Stage 2 practices into the points of interface with customers and suppliers. The company has identified strategic customers and suppliers, as well as the key information it needs from them in order to support its business process. It uses joint service agreements and scorecard practices, and takes corrective actions when performance falls below expectations.

Stage 4. Customers and suppliers work strategically to define a mutually beneficial strategy and set real-time performance targets. IT and e-business solutions now automate the integration of business processes across these enterprises.

Looking specifically at plan practices, two thirds of the respondents' supply chain organizations are at Stage 2 and above. This indicates that they have embraced internal integration across functional boundaries in their supply chain planning and demand/supply balancing processes. This relatively advanced state of plan practices ensures that a well-structured and organized process is in place to support both strategic and tactical decisions. Few companies have established the Stage 3 practices of true collaborative planning, whereby demand and supply information is captured directly from customers and the coming goals would be shown. Product flexibility is an important idea behind best practices in understanding the above mentioned manufacturing flexibility, where the main idea is a possible measure indicator or the time needed to switch from one part mix to another. The questionnaire for this stands as: the average cycle time for new products and their frequency of development and reaching it through a proper production planning and control systems.

To be most effective, the choice of process technology which is a best practice is contingent on the strategy choices the factory has made. Thus, the choice of manufacturing and technological processes reflects the degree to which the firm will compete through manufacturing. Performance should be related to the match of the choices of manufacturing and technological processes for the chosen strategy. The match is the linkage between the first two manufacturing strategy paradigms: competing through manufacturing and strategic choices in manufacturing. The final domain, structure, includes the definition and classification of tasks to be performed into organizational units. It also includes the mechanisms to integrate and co-ordinate individuals, groups, and organizational units. Giving a strategy focus to the manufacturing strategy the main idea in the cluster relationship is, the contingent nature of the characteristics found within each of the clusters represents the role of strategic choices in defining and formulating manufacturing strategy, the second manufacturing strategy paradigm. Each cluster is a set of contingent, or linked, practices which should be selected together for maximum effectiveness. This is consistent with the process choice model and the general findings in the configurations literature. The composition of the individual clusters is worthy of discussion. The strategic focus cluster includes strategic adaptation, technological adaptation, and management practices. A review of the items comprising these scales reveals how they address the choices of a set of integrated issues related to product features and the methods to produce it. Strategic adaptation addresses the nature, scope and the strategic linkage between manufacturing and the business unit. The choice of strategy requires subsequent decisions about the types of structure, technology and location of decision-making authority. Technological adaptation and management practices items in this cluster reflect these choices. This cluster shows how manufacturing strategy is consistent with the configurations in literature, which has been much broader than the process choice model in its analysis, but highlights the contingent nature of decisions which flow from product/market decisions. Indirectly this indicates how manufacturing can be a source of competitive advantage. The second cluster, operations management system, represents factory floor operations and management. This cluster is a detailed extension of the contingency nature of decisions on the factory floor. The scales included production system, working on the shop floor and production control system, including traditional factory floor practices as well as more recent concepts such as rewards systems, group problem solving and supervisory facilitation rather than directing behavior.

There are other types of benchmarking activities like flexibility which are related to design, mix and volume which are related to the key performance areas like mobility, uniformity, range. The other major best practice is reaction time which is totally related to order shipment time, order preparation time, order manufacturing time, order processing time and order transmittal time. The other best practice is quality which consists of durability, design, conformance, serviceability, quality improvement, perceived quality, reliability which are at last supported by the last best practice i.e. return on assets consisting of net sales and net profits. The performance measures like how much can companies improve its performance? And how to dramatically improve the performance and ways used by performance measures that would be used to compare each process? The control patterns are: expense budgets based on strong negotiation processes and focused on resource allocation more than on the definition of performance targets; expense controls based on a system of bureaucratic authorizations; managerial controls based on the limits of expenses set by budget. A performance indicator is effective when it activates behaviors that are congruent with the generation of value for a client internal or external. Since business processes represent a logical bridge between the value generation strategy and the daily operations of the firm, performance measures should be process based. While business processes have clearly identifiable external customers, supporting processes, such as accounting processes, contribute only indirectly to value generation for final clients: they release functionalities to business processes that should be considered as their internal clients. Connecting performance measurement of accounting activities to value generation implies two logical shifts in performance measurement methodology:

(1) Changing the point of view: from organizational units the accounting department to organizational processes to the accounting processes.

(2) Expanding the traditional cost based perspective: from a uni-dimensional perspective typically cost-based, to a multi-dimensional perspective considering process cycle time and quality of services as well as cost.

Defining cycle time in that way is simply not congruent with a value generation strategy: cycle time reduction could be declared as an improvement in performance while, in effect, it would only imply a waste of resources. The best practice enablers are proposed in four steps: The first step is defining the boundaries of the process. The complete a/r process is a sequential flow of activities, starting from the credit approval for a new customer (or a credit check for a new order issued by an old customer) and ending with the cash payment from the client or with the write-off of the credit. The choice for such a wide definition of the boundaries is intended to offer a real multi-dimensional performance evaluation of the process. In fact, this is a real inter-functional process, in which many business units and departments, e.g. MIS, finance, distribution, etc., are involved at different times. The second step is segmenting the process. Two basic alternatives are available for segmentation: segmenting into phases or by macro-activities. Segmenting the process into phases emphasizes the process flow and helps in the identification of the connections between activities that are most critical for time and quality. Segmenting the process by macro-activities means to aggregate tasks that are homogeneous for technology applications and independently of their sequence. This helps to identify areas for improvement in quality and cost. The third step for identifying best practices is the definition of the performance vector. In the case of the a/r process, the performance may be measured in two ways. The first alternative emphasizes the cost/productivity dimension: typical measures include the average cost per order processed (critical for manufacturing firms) or per invoice issued (critical for engineering and construction companies) or per active customer (critical for evaluating the selling strategy of the firm). The second alternative stresses the financial flows connected to the a/r process: typical measures include DSO (days of sales outstanding) and credit aging. The fourth factor deals with identifying the flaws and defects in the accounting principles. Moreover the adoption of best practices can be limited (or favored) by company choices about vertical integration, plant localization, and strategic connections among businesses, clients and suppliers. Assume again that a company gets the authorization to self invoice on behalf of its suppliers. Another series of obstacles might appear, such as the dispersion of the company's supply system among a large number of small suppliers competing one against the other, each of them against the client, to offer the lowest price. Self-invoicing practices are more efficient and effective when there is a small number of suppliers and when there is close co-operation with suppliers.

Similarly like best practices balanced scorecard is a technique that helps grow the internal resources of a business. All the ideas developed on are outside the very creative balanced scorecard techniques developed by Kaplan & Norton. The main use of balanced scorecard in practice in organizations is decision-making support at top-management or operational level; but the performance measurements, such as their short-term and historical focus, bias towards financial performance and information overload, exhibit many shortcomings. Balance scorecard helped to reduce tensions between management and professionals. The scorecard was used for horizontal, rather than top-down, communication. The horizontal communication facilitated the achievement of goal congruence. Balanced scorecard is a more effective means of understanding and promoting value creation, productivity and profitability than traditional management control methods. Additionally, it is not obvious that the BSC is potentially better at bridging the abstraction gap and surmounting the mental barrier between top management and employees. The balanced scorecard creates a static-ism that tends to struggle with the challenges of highly competitive and changing business world. Within the balanced scorecard approach, a centrally defined strategy is translated into certain measures that align all company activities to achieving its balanced scorecard goals. In consequence, the optimal implementation of a BSC leads to a high level of uniformity and goal orientation. This increases and possibly maximizes the focus on the given goal; but it limits any further activities and initiatives that might go beyond the originally set targets. Static-ism therefore results in a high level of entropy, namely loss of a significant amount of energy that is not used within an organization. In such an aligned organization employees might have a clear perception of their job, e.g. in the objectives and achievements of BSC metrics, but they will do little more than only achieving just these. The BSC recognizes the temporal dimension of strategy. This is reflected in the inclusion of the scorecard to help translate the strategy map into a set of performance indicators that spans a period of time. The relationship depicted in the strategy map is used as the basis for the developing the scorecard. The scorecard spells out the short-tem and long-term performance indicators of the strategy. The second generation BSC emphasized the cause-and-effect relationships between measures and strategic objectives. It became a strategic management tool, usually utilizing a strategy map to illustrate the linkage between measures and strategies. In contrast there is a view in the literature that the key contribution of second-generation BSC was the formal linkage of strategic management with performance management. According to the third generation BSC, it is about developing strategic control systems by incorporating destination statements and optionally two perspective strategic linkage models. They used activity and outcome perspectives to instead of the traditional four perspectives suggested that the third generation BSC was the second generation but adding action plans/targets and linked to incentives. The balanced scorecard integrates measures taken from the traditional financial perspective as lagging indicators of performance with non-financial measures as leading indicators of financial performance. These leading indicators are taken from three additional perspectives: the customer, the internal business and the innovation and learning perspective.

The balanced scorecard scrutinizes and assesses groups of functions and supports the adaptation of the identity and mission of the organization. The functions involved are policy, intelligence, and control. Together, intelligence and control generate relevant and feasible plans for innovations that either elaborate on or expand the organization's current identity and mission. Scanning relevant developments in the environment of the organization and relating them to potentials for change is crucial for the generation of these plans. The policy function balances and facilitates this discussion between intelligence and control and consolidates its results in plans for innovation. If we look at the BSC, we see that it allows for a cascade of strategy maps put to work by the four management processes and the five management principles. The maps support the interactive unfolding of the organizations' vision and mission into scorecards. These scorecards are consisting of objectives, targets, measures, and activities. Moreover, the maps, processes, and principles, allow for monitoring, controlling, and adapting these objectives and targets. The four perspectives constituting the core of the maps and their relations as leading and lagging indicators provide a strong framework for hypotheses in terms of cause and effect relationships as how to realize the objectives and targets in the four perspectives. In combination with the management processes and principles, these hypotheses can be realized and adapted, thus focusing the organization on its strategy. The BSC had been expanded into an organization-wide strategic management system, which enabled managers to break down high-level business strategy into agreed operational initiatives for constituent business units within each of the four performance categories of finance, customer, process and learning and growth. An initial intention of the BSC was to reduce managerial bias towards financial measures in performance evaluation. Although this holistic approach is in one sense more in keeping with the principles of the BSC. It was viewed as a single entity; it introduced additional subjectivity and the potential for further cognitive bias, which served to cloud the true decision-making methodology. These varieties of decision methodologies used implies that training of evaluators in the use of the BSC might increase the consistency of evaluation techniques, but would still not address such issues as current strategy understanding, processing limitations, other existing cognitive biases or being comfortable working with figures.

The BSC reduces the pressure on managers to find and install the perfect structure. Whatever structural choice they make, when managers apply the enterprise scorecard to functions and units, these units become better aligned with each other and to the enterprise. The process starts by applying the BSC at a high level, and letting the strategic objectives and themes cascade down to lower levels, where they are interpreted and customized to the specific situation faced by lower level organizational units. This is the process by which decentralized units become aligned with each other and to the corporate strategy. Executives can also use the comprehensive performance measures on the unit's BSC to assign clear responsibility and accountability for local and global performance. By using the BSC as the primary organizational system, decentralized unit scorecards reflect both the specific competencies and strategies of the unit for local success, and also how each unit links with other units and the enterprise as a whole for the enterprise strategy to be achieved. The S skills are the other skills that are captured in the internal process perspective of the BSC, where organizations identify those operating, customer management, innovation, and regulatory and social processes that create and deliver the distinctive customer value proposition; and lead to achieving the financial perspective's productivity objectives. In developing the strategic objectives and measures for the BSC internal perspective, the organization enhances its skills by aligning human, information, and organization capital from the learning and growth perspective to critical processes, and also by selecting strategic initiatives specifically aimed at enhancing the performance of the internal processes. Thus the BSC becomes a powerful tool for aligning skills, the organizational competencies, to strategy.

The transition from second-generation to third-generation balanced scorecard designs, although in terms of design elements less significant than the earlier transition, represents a significant change in the approach to balanced scorecard design activity. The adoption of third-generation balanced scorecard designs has been particularly helpful in supporting the development of multiple balanced scorecards within complex organizations. Corporate performance management software systems have been presented by some as a solution to part of this problem by making it economic for large volumes of detailed information about activities and performance of the organization. These are to be collated and assessed centrally and a key feature of such offerings is the ability to drill down into information recursively to get to the root cause of performance anomaly. However, the information asymmetry viewpoint challenges the utility of such activity, as the software provides at best only a partial solution to the asymmetry problem like you may have more data, but not necessarily any more information about the local context that is necessary to make this data useful. Similarly more complex alternatives to balanced scorecard do not openly address the informational issues presented by this increase in complexity. It has been shown that one development of third-generation balanced scorecards has been to support alternative management models that tolerate or accommodate the information asymmetry issue through facilitation of the concise articulation and communications of key data, and through facilitating the identification communication criticalities in an organization's hierarchy.

Moreover taking the case of balanced scorecard and R&D, BSC needs to be implemented in R&D because of the scale of changes that have taken place in recent years. The rate of growth in the size and scope of R&D departments has been spectacular and rapid, to the extent that problems of visibility are being generated. Managers feel that the basic decisions that were taken relatively easily years ago have now become extraordinarily difficult. In the case of validation of BSC for R&D, each group of items or empirical indicators represents one of five strategic dimensions for the companies: these dimensions cover financial results, customers, innovation, internal processes, and learning and growth. The BSC model for R&D developed in this study has been subject to testing with recognized experts in management and in R&D; however, in a future study, already in progress, the reliability and the validity of both the construct and the criterion of the scale will be studied in depth. Regardless of future possibilities, this first phase of content validity has enabled us to put forward a proposal in respect of those indicators that best define the factors related to organizational effectiveness in the achievement of the strategic objectives set by companies, and to inter-relate them and group them under five broad perspectives of the BSC. The relationship between strategic planning and BSC is important, since they can be considered as complementary tools. The function of the BSC is in translating the strategic guidelines described in strategic planning in such a way that everyone in the organization can understand them so that the strategy can be implemented, monitored and overseen in the long-term. The activity of planning is also very important in the effort to producing quality; this has received the greatest attention in the current model of quality management, both because of its natural importance and also because of having been considered of little relevance until quite recently. According to the perspectives of the BSC in the quality environments, it permits better working of the action plan. Like what can be determined, even in the planning process, who will be the people responsible for drawing up and supplying the necessary information, where resources may be necessary. So those people may be involved in the process of implementing actions for improvements, where sectors should display a greater link based on the objectives to be achieved. Intertwining the BSC with the quality environments gives the evidence for quality being contained in all company's areas, thus giving greater support so then the objectives of the main stakeholders may, in fact, be reached. According to the perspectives of the BSC, there should be an evaluation of the results obtained so as to compare these with parameters which have been used by chance, or with targets which may have been defined, so seeking not just the targets achievement but also the reflection of these results in the other perspectives considered. It is also important to evaluate the impact which alterations in one company's area affect the other ones and to know how the degree adjustment will be needed.

CHAPTER 5

Bringing together new and breakthrough technology by utilizing the internal resources
(BY BAISHAM CHATTERJEE)

New technology often depends on R&D and competitive strengths for translating its R&D efforts and achieving superior R&D productivity. There is necessity to make advancements on the core subjects like chemistry, information technology, electronics and material sciences. The emergence of Java as an internet scripting language has brought big changes to push forward demands in the future. Research organizations form a key advantage to generate technological options through experimental facilities and teamwork. This research outlines helps in developing through beliefs, effort and creation revolutionary and evolutionary R&D projects that provides the technological potentiality to provide the highest technology yield. Different projects in different countries can be all measured in a diagram where countries should have following qualities:

- Potentiality to take up new projects through revolutionary technology can be assessed.
- Moreover the variability of modern business and paradigm shift in innovation across different countries through the 10 year analysis graph is another phenomenon.

- Core competence of a nation and emerging technology in other countries is also very important to measure.
- Moreover it is also very important to analyze big corporate, SMEs, work culture, university development in the future.
- Another important country related idea is looking at the economic resources and the country's ability to use the new technology and previous experience of using the technology.

In these ways 100s of charts and graphs can be prepared that can put forward the competitive ability of a nation in new technology.

Virtual businesses originate from the impact of modern computers and workstations that lead to providing an advantage in bringing changes in responsiveness, incentives and communications. E-marketplaces provide high powered market based incentive to provide attractive bonuses and stock options. It is stated that as incentives to take risks increase integrated corporations are formed. This helps in bringing centralization through alliances and joint ventures that ultimately lead to building corporations. This has a knowhow of its inner risks, culture, beliefs, motivation and research that would change the way the firm communicates through its other fields of study and departments. If someone considers as to how technology standards emerge market participants weigh many competing technologies. In this case the expert and the futurist work together to bring innovation, marketing, behavioral science and statistics together. Virtual firms generally tend to die out due to too much centralization. Virtual companies that have survived and prospered have outsourced everything and have searched for markets that are more welcome to learn and collaborate through internal talent and thus can provide the opportunity to lessen ones burden.

With new capabilities that are approved through patent rights or core competence as in electrical equipment, electronics, machinery building firm, carries a high sense of disruptive as well as increase high specifications. This has given high yields and controlled processes that gave a significant cost advantage over competitors. All the above factors help in commercializing complex products more easily and penetrate markets more rapidly than before.

When a new product is introduced as a suitable competitive advantage it takes some time for manufacturing performance in terms of costs and productivity of labor and yields. This to a large degree is also known as ramp up which is also known as a means of process development. The faster a company can ramp up production of a new product the more quickly it can earn significant revenues and enables companies to penetrate the market quickly, gain broad market acceptance and begin to accumulate experience through high volume production.

Production process determines many specifics which are largely dependant on the application of superior process capability early in the product development process, with superior process capability and a very composite link between process technology and product characteristics. There are increasingly being a shorter innovation cycle due to the high innovation capacity, increasingly hard to manufacture product designs, fragmented markets and growing technological parity. The most talked about among this is an increasingly hard to manufacture product design that creates high imperatives for manufacturing that requires rebuilding on simple to use manufacturing processes and that ultimately removes the challenges. Historical revenues is another factor of study where fixed development and revenues with greater price pressures and shorter cycles helps in developing ideas on how to shift by contributing and thus map all the developments before the firm emerges as a leader. It can be said that the effects of process development are cumulative and the earlier a company undertakes process development the greater the total financial return. Process development is thought to be so much into the financial life of a product that it is best to avoid scarce technical resources in a project that may never make it to market and it is very important to make a joint development idea between process development and a launch schedule. Pilot production is another important schedule which has been playing an important role in producing enough materials so that the appropriate clinical tests of the product could be conducted.

There are various forms like the integration team that investigates the impact of various technical choices on the design of the product and manufacturing system. There are various measures of the existing capabilities of research and manufacturing capabilities that come together in managing imbalances and manage the process improvement as well as the technological development that removes the majority of the uncertainty in any high-tech business. There is always a strong product concept in building the exploration and integration ideas behind new idea creation and utilizing it for any future calculations and building the system that can help create any form new ideas into visible form through the very personal technology that is created. R&D time is an important phenomenon that can bring about true research ideas through the time through which concepts can be approved and projects can be started and developed. This is based on the calculative resource allocation and the number of people and their skills in taking up that work. Resources used in certain product generation gaps as well as managing integration teams provide the understanding of performance gap between traditional and system focused companies. Competitive pressures to keep up with increasing technological innovations led the idea to bring the resources they sunk into technology development.

It is very necessary to establish R&D sites. R&D sites may be classified into two parts home-base augmenting laboratory site and home-base-exploiting laboratory site. Creating new knowledge through local manufacturing and marketing as well as transferring knowledge from the company's home base provide the key base to new system based and consequence based technology that can proceed very fast. Dynamics of R&D can be well achieved through encouraging employees and exchanging researchers with local university laboratories. There is said to be many magic formulas for success, there are also magic formulas for rapid new-product definition, because new platform products are developed to serve the future need of customers. Companies are at high risk when they work with new companies and improved technology or architecture. It can be found that products emerged faster and made more of an impact on the market when product teams included the company's most knowledgeable engineers and marketing strategists from the outset. It can thus be moreover seen that whether it may be technology integration or organizing for innovation, there are different ideas that can be used to create various R&D capabilities abroad as well as defining next-generation products. These are a collection of various ideas and thoughts for creating next generation products.

This is all a brief structure to say what is possible in the modern world and what are the difficulties but to gather ideas in order to reassess a company's goals and reach the structure to understand an organization's problems. There are a lot of problems involved in individual organization's and its goals and provide and bring together opportunities that are scarce as well as work on organizational mission and time constraint. This gives a clear picture of where a firm should go and how its future can be shaped.

Technological perceptions are very important because technological forecasts are playing a progressively more important role in the selection and implementation of applicable technologies to aid in initiating the first mover advantage in the market place. Methods used in understanding future of data storage were patent trend analysis, bibliometric analysis, and technology cycle time and growth curve. Patent databases are valuable for analyzing a firm, industry, technological competitiveness and technology trend analysis. Firms and researchers can utilize patent analysis to assist in strategic planning efforts and assess specific innovation researched and developed all around the world. Technological scope and innovation capabilities of a firm as well as inter-industrial knowledge flows are necessary to be understood. The Technology cycle time (TCT) was applied to measure the technological innovation progress for these forecasting technologies. Technological development is an important criterion behind this.

The developer should interpret information from the marketer, which is required to fulfill this task. Learning and using opportunities are very important with factors like direct communication and design task. Connections with internal IT infrastructure and external vendors were well-established. Connectivity is a serious issue that leads to understanding internal systems, particularly those that have been in place a long time. Connections to the mobile platform are critical, where convergence between platforms is required. Devising and ordering business options is another important priority which leads to assessing and selecting business options, specifying and designing business opportunities which ultimately leads to evaluating business opportunities which is subject to formal life cycle and intermediate feedback. These types of operations lead to proper functioning and intensive ideas behind using the communication channel. As of course looking at not only the different new technologies that are reaching the market or evolving through reengineering, the third source of market inefficiency in technology creation lies in its uncertainty. This shows the difficulties of estimating the technical or commercial returns of an innovation. Innovation related risks can be accurately computed and used in actuarial calculations of expected costs and benefits are fanciful in the extreme. Probability calculus suggests the ideal way of analysis. The unpredictable delay between knowledge creation and application is very important. One immediate consequence of this is asymmetric information between firms and potential suppliers of capital between the R&D managers and the firm's board of directors. Potential lenders cannot judge accurately the claims of technical personnel. This is when adverse selection and moral hazards are the consequences. This is when it is difficult to ensure technology projects. In this way they can pull the risks between a portfolio of projects and it helps us to better understand the pressures towards more collaborative work in R&D and towards mergers & acquisitions between technologies based companies.

The new growth theory as suggested depends on demonstrating how the structure of an imperfectly competitive industry depends on the nature of market demand and on the relationship between advances in technology and R&D inputs necessary to achieve those advances.

Push technology is a new technology that has been recently introduced in the market. Push technology allows an organization to build a decision support system in which updated information is pushed to the desktops of all relevant managers within an intranet/ extranet. It consists of a lot of information sharing where sales and order information recorded daily by field salespeople is pushed to the desktops of relevant employees. Secondly the webcasting of customer, requests for samples facilitates the management of inventories. 12 years back there was around 16 vendors offering push technology and out of that, the very new ones are Lancôme inc, Wayfarer Communications are upcoming vendors that contribute through this technology. Push technology helps to subscribe to channels which monitor the price changes of competitors and disseminate instantly their competitive pricing responses to consumers. Consequently, push technology is likely to reduce the effectiveness of the tactic-endemic in industries such as air transportation and gasoline—of attempting to steal volume from competitors through short-term price reductions.

New technology can be created, but there are various factors involved in managing new technology and knowledge with the purpose of evaluating and managing the uncertainties the organization faces as it creates and protects enterprise value. Enterprise-wide view of an organization means 3 things and that control cultures facilitates information flows up and down and rarely horizontally. Secondly an enterprise view should reward and recognize decisions, an approach that can reduce the effect of risks through incentives. Thirdly risk management goals, objectives, policies and processes must be consistent with the view of the organization management; operating units understand the circumstances to help in the success.

Risk factors somewhat help us in understanding the circumstances and prepare for the circumstances, but there are very new pricing methods that are also to be introduced. Electronic commerce and communications, and the creative use of the internet now provide brand managers with the ability to quickly gauge the potential for new product success, at speeds that far exceed traditional market research techniques. The tools of the new economy enable test marketing of new product prices; segment based pricing and instantaneous competitive price benchmarking. The emerging technologies are also enabling consumers to better visualize and experience new products prior to their introduction to the marketplace. The electronic delivery has made it possible for products to be created, modified and delivered in record times. There are many internal qualities involved in organizations that can use internal communication systems or intranets to identify emerging attributes or factors that may be developed into distinctive competencies. Internal technology can keep participants abreast of meetings, results, and implementation plans. Communicating via intranets, within companies and other electronic modes, individuals may be more likely to offer appropriately critical evaluations of strategic options.

Moreover, there has been several dimensions to effective performance which are as advances in computing and globalization that has helped in creating new systems that enhance performance that can help in finding new markets. It is also very important to understand the buyer patterns as well as understand and integrating such diverse disciplines as technology, services and graphics. It is also a necessity to move into market first and use their data/knowledge about what a customer has done in the past to sell more in the future. The internet fore mostly has helped in finding physical aspects for their products to capitalize and understand their potentiality. Internet firms that emphasize innovation and rapid response to change may be best positioned for recognizing and identifying new opportunities and ideas for their business. Identifying new opportunities can be critical in initiating innovation and change associated with improvements in products/services, technological capabilities and seeking alternative markets and opportunities. Internet firms that adopt a prospector strategy continually search the marketplace for new products, services and technologies. It is important to understand new product and market opportunities where organizations with a prospectors strategy are the creators of change in their industry and business which all depend on high level environment scanning and long range forecasting. Proactive entrepreneurs were more likely to introduce new product/ service offerings and build e-commerce solutions/applications to satisfy the needs of both the organizations as well as their customers. EDI (electronic data interchange) and EFD (electronic fund transfer) are the key resources that help in developing features that lead to developing and creating the channel and customer relations for a business. A very simple idea on the success of new technology can be found in the usage of mobile phones and their applications.

It is not only that one can bring a new technology, but it is also very important to commercialize a new technology. It is very important to validate commercialization and then realistically assess utility of the technology. This helps in understanding the cost measures through the various actions taken by the project teams' ideas, their core prospects and understanding the cost beliefs.

Emerging industries can be created when new products enter the marketplace as a result of technological advancements. It is also true that buyers' technology concerns are informational and their ideas are being formed and understood.

CHAPTER 6

Enterprise resource planning: bringing out new change in internal development
(BY BAISHAM CHATTERJEE)

Enterprise resource planning helps in generating real time data. It enables linkages across functional areas and divisions and thereby, promoting the integration of information across all of the units of a business at one time. SAP R/3 is a type of ERP system that is used in any kind of development in businesses for development in the modern world. It helps in developing real time data and develops linkages across functional areas and divisions and helps in integrating information. The main strengths of ERP are: either information or data that are transmitted and are translated into the respective languages of the global installations. Again, this feature only serves to strengthen subsequent decision making. Utilizing a common software system, utilized by both corporate headquarters and respective global locations, certainly puts into motion this capability. Moreover it is found that ERP is related to BPR in many ways where the most important functions to manage are the resistivity to change; lack of senior executive support; lack of cross-functional project teams and staff and neglecting employee values and beliefs. These are different errors in bringing system failure or malfunction of ERP. Only a BPR and TQM together can support an ERP. The essence of a complete ERP system is to automate business processes, share common data across the organization but most importantly, to produce real-time data. ERP systems offer distinct advantages to companies adopting them, improving the decision-making process via the provision of appropriate and timely information. Further, improved planning and control of operations are also derived from ERP applications. ERP systems of companies and accounting processes help in developing business processes. Transactions under ERP systems are treated as part of the inter-linked processes that constitute the business in its entirety. Such systems allow companies adopting them to automate and integrate business processes, share data across departments and produce and access information in real-time environment. For example, entering a client order to the system would be sufficient to update all its relevant parts, such as stock levels, general ledger and logistics. In essence, a complete ERP system would incorporate a number of modules relating to not only the traditional accounting information system, but also stock control, MRP and logistics. Additional dimensions of an ERP system might involve EDI systems, and e-commerce. An important problem is the integration of ERP systems with the existing legacy systems. Overall, their core advantage, i.e. the interdependencies involved, may also constitute an important limitation resulting in data errors and business interruptions. Substantial cost and time overruns, organization problems such as employee resistance also appear to be important barriers for their success.

In the system of integrated software controlling logistics, cost accounting and other features relevant to production systems, these are recognized as the ERP tools. They develop standardized packages that are characterized by platform independence and industrial situations. Bill of materials is a listing under ERP where routings and recipes are crucial examples which has an item and parent that is replaced by a formula which is focused on substances rather than parts and that have a more flexible nature. Moreover looking at the power of ERP in understanding Environmental information system which is available as an isolated software tool, like waste administration packages. Integration and co-operation of EIS and ERP requires an emphasis on their common base: the physical flow model. The business blueprint within SAP R/3 concentrates on four key areas necessary for understanding business: events, functions, organization and communication. Events are the actual driving force behind every business process, prompting one or more activities to take place. SAP R/3 is a modular designed and fully integrated application incorporating among others product supply, manufacturing, logistics, procurement, sales, operations planning and finance. ERP systems have helped to upgrade core business activities, importantly; they have not solved many of the underlying business structures and process problems. There are many results regarding ERP and other organizational issues. They point out that rapid organizational change and increased complexity in new product development will increase the organizational demands on existing and proposed ERP systems. There are also many gaps formed in these ERP systems that should be understood by standardizing ERP models. Standardization of ERP models as base for common way of working gives flexibility for organizational changes. Company-wide business and ERP templates bring about standardization which makes these gaps less vendor profitable and opens for customers to change vendor and outsource ERP maintenance. ERP has many phases of implementation where the major qualities are pre-implementation phase where the planning process has to be identified, and before the implementation the expectations for benefits realization, magnitude of change, change ownership, process redesign and functionality delivery options has to be understood. A system build determines the software components of the ERP system and how these components interact with each other. Business processes are analyzed to understand the current conditions and functional and technical requirements to understand system build needs. Later the analysis phase forms the foundation for process redesign, and system build and change management. Business processes, functional and technical requirements should be focused upon. Customer influence is developed through user influence, team influence and organizational influence. Process flows are through the understanding of finance which is composed of ideas from purchase order process, stock flows in and out and sales invoice. Other fundamental areas to deal with are HR, SCM, CRM, SRM, BI etc. Moreover while looking at ERP installations; they often help small and midsize manufacturers to improve their strategic and competitive capabilities. ERP implementation has sometimes resulted in total abandonment, and small manufacturers often lack financial resources in integrating ERP systems IS staff levels and extensive IT training and development. ERP systems are comprised of a suite of software modules, with each module typically responsible for gathering and processing information for a separate business function, or a group of separate business functions. ERP software modules may include accounting, master scheduling, material planning, inventory, forecasting, finite scheduling, distribution planning and others. A typical ERP system integrates all of the company's functions by allowing the modules to share and transfer information freely. ERP systems is a type of best business practice that enables the implementation of these practices with a view towards enhancing productivity and in empowering the customer to modify the implemented business process for specified needs. There is assumed to be many migration tasks in ERP implementation where the workbench helps the user to define appropriate strategies, decide architectural configurations and select a software package. They are developed based on components rather than modules and will be designed for incremental migration rather than massive reengineering. The process is on managing dynamic rather then static configuration which in turn requires managing multiple sourcing and partnership relationships.

ERP systems have further developed the scope of standard software systems providing systems to support all business functions. Advantage should be taken on the standardization and integration of business and decision making. It is important to have these kinds of integrated systems that enable implementation of best practices and triggers business process re-engineering and supports quality drives. But this implementation is often costly and complex and needs a multi-dimensional construct. ERP systems is a construct that helps in developing decision making for the support of the organization as well as understanding planning and control. All this evaluation is formed through the collection of data that eventually brings profitability. The first stage of ERP implementation encourages users to experiment with the system, to test its boundaries and capabilities to establish the ERP system. Only then can users begin to move into the second phase of learning to improve the system. It is important to understand new reports, new data, new communications protocols that would enable them to manage the resources at their disposal.

CRM and ECM are concepts that represent a consistent application of best practices in the business process making optimal use of people, technology and other ideas. Resource allocation necessary to implement ECM/CRM is justified by the fact that research has shown consistent increases in shareholder value that have been a characteristic of companies that have made it a cornerstone of their strategy. CRM/ECM manages service encounters and process elements where the customer makes a determination of satisfaction with any respect of the organizations offering. Positive outcomes build loyalty on the part of the customer and cement relationships. Another important idea to deal with is the customer relationship life cycle support system that will allow delivery of customized sales: efficient delivery and service support, and maintain high overall customer service perceptions. Improvement in business processes are formed through the correlation between orders, their opportunity and the delivery. In this phase the relationship is between sales, design, procurement, production, inspection and dispatch. The integration between CAD/CAM systems and ERP systems cannot produce maximum results unless the data collection (interchange) is appropriate for the task. Traditional systems rely on a hierarchical, top-down model that conforms to the host hardware-processing environment, with extensive data interchange networks submitting transactions over a dedicated network. In these systems, transaction verification must be performed at the host level, leading to constant disruptions and delays for users. Therefore data interchange becomes a roadblock to effective CAD/ERP integration. With continual interruptions to the host process, the modern ERP system can be rendered ineffective. These problems and the inherent complexity of integration have led a movement of hardware-driven improvement in data interchange systems (DIS). The result is that modern DIS in next generation manufacturer will incorporate client/server architecture, extensive connectivity options, multi-platform computation and modular construction. This DIS must be deployed in accordance with the connectivity requirement of both the CAD/CAM and ERP systems. It should be remembered that the CAD/CAM connectivity requirement is normally for LAN connectivity while the ERP system requires wide area networks (WAN) connectivity. In general, this is because CAD/CAM systems run from LAN platforms while ERP systems run from WAN platforms. ERP is a major positive focus in manufacturing, because it is a fully designed solution that is achieved with open systems by selecting strongly featured departmental application systems and then integrating them. Neither enterprise wide computing nor strategic alignment concepts support this notion. One of the technology factors that really differentiates the enterprise-wide system from other clustered applications and develop an enterprise design model with two core building blocks upon which next generation information systems are built. The benefits of ERP range from price/performance improvements of the technology to the enterprise solution that finally ties controls and engineering directly and strategically to business systems. ERP and related systems must interconnect with manufacturing processes to maximize the use of resources. Automation and control software, MES, plant portal and information systems, and other software aim to provide the interconnectivity and two-way information flow needed. ERP solutions use technology to address business issues, at the same time striving to keep technology transparent for the users. Users do not need to learn more about bits and bytes but they need to know how operational and long-term business issues could be effectively addressed with technology, with a user-friendly interface.

The typical lifecycle of an ERP system development involves more steps than a software project lifecycle. The reason for this is that an ERP is a complete flexible business solution while a software solution is more of a limited and fixed solution to a specific business need. The software project life cycle provides various ideas including analysis, design, implementation, testing and debugging and documenting. ERP includes six elements that touch business, application and technology that has ideas of business domain, the functions addressed within that domain, the kinds of processes required by those functions, the systems architecture that supports those processes and looking at data and its architecture. The existing ERP systems has a discontinuity view where exchanges drive the supply chain integration and the monolithic ERP is replaced by fragmented systems and that ERP only is needed in the intermediating marketplace hub. At least in the short term there are no signs that the enterprise is giving up on their corporate ERP strategy but the network perspective of the ERP transformation is important. ERP ensures lateral transfer of information to the different departments. Imagine the different measures as to what customer order processing would be without an integrated system. Various departments in the organization can track the status of a customers order and quickly relay the information to the customer, thus offering better service to the customer. This approach could also be cost effective, since it eliminates the need for an individual database in each department. If ERP is properly implemented it cuts down on bureaucratic structures and would decrease cost, improve efficiency and productivity and facilitate information acquisition. Internet environment and information plays a major role in this integration. The customer centric approach is the driving force of an organizations competitive sustainability. It is a tool for providing value, high quality and productivity at reasonable cost and achieves an organizations mission. Along with ERP a new structure will emerge that will encourage matrix-oriented organization. These teams become the brain of the organization since they manage the inter-relationship between the different functional departments and may redesign the role of people and decision makers. ERP systems make large enterprises rely on information technology more than ever. Automate routine process in areas such as accounting, inventory control, and procurement that accomplishes organizational accessing through automatic updating of the transaction data. ERP connects various functions of the organization in an integrated fashion. It improves the responsiveness to customer needs and delivers products to market quickly through compressed cycle times. ERP system makes data available in real time and hence allows for a more comprehensive and unified data management.

The ERP and BI framework can bring out the following benefits through enabling finance personnel to generate revenue/expense reports quickly, allowing the controller to recognize corporate cash flow in real time; sharing sales information with the management. This allows for making better corporate decisions based on a macro view of the business; facilitating a company to perform interdepartmental collaboration; improving accounts payable and vendor relation management; enabling sales force management; improving profitability by analyzing transactions data and forecasting business trends; improving customer relations through in-depth sales data mining; providing online access to data, which saving access time; and reducing time to generate regular reports. The issues of control and coordination change slightly with regard to ERP systems. To achieve the benefits from an ERP implementation, organizations usually standardize their business processes. This is part of the debate with regard to MNCs at which point it is beneficial to have variations in processes in different globally dispersed locations. One issue regarding ERP systems is clear; to achieve any of the benefits the organization must establish process and data standards including nomenclature, file and field sizes, and common part numbering where applicable. This in itself is a base level of control which must be established. Accompanying these standards is usually a series of procedural standards as well which govern how the system is used and what processes are integrated. Economic and strategic justifications of an ERP project prior to installation are based on involving enormous investment and risk, and creating values and modules that provide the yardsticks for measurement. Ex—when sales and marketing modules are integrated in concert with the financial reporting function. Management is able to make important decisions based on a detailed understanding of product and customer profitability rather than instinct. ERP enabled business processes are designed to evolve and grow in power for those organizations that take time and effort to grow with them. The first wave implementation hinges on companies that can achieve a smoother and tighter fit between their business processes and ERP systems. Moreover a new form to deal with is the APS that derive sophisticated mathematical algorithms to model and analyze the supply chains constraints to develop plans that provide optimal and sub-optimal solutions.

The six business processes of ERP are quote to cash, procure to pay, plan to perform, manufacturing operations, product life cycle and financial management. They derive ideas by identifying the qualified customers with needs, apply company's products to address the needs and conclude with customer payment. Managing finances, order flow, managing inventory, managing production and product life-cycle helps in bringing out the key business flows. Effective ERP implementation requires putting in place appropriate managerial interventions as part of the implementation. For example, we consider the institution of training programs and communication mechanisms to facilitate understanding of the technology as part of managerial interventions. Managers design interventions, and usually the appropriateness of these interventions will most often be based on the perceptions of managers. For example, managers, because they are part of project team leadership, might think that an appropriate means of communicating about the project will be the minutes of project team meetings that are disseminated to employees. On the other hand, employees not being part of project teams might consider minutes of team meetings as being inadequate and might desire more effective communication mechanisms. Thus, understanding the differences in perceptions will be important to designing appropriate interventions that might lead to effective ERP implementation. ERP systems are designed to provide one common source of data. They are designed as integrated information systems that eliminate multiple sources of data, eliminate multiple data entries and provide more accurate and timely data. The data integration means that an incorrect entry by, say a customer service employee, can have simultaneous organization-wide implications. This can be a source of anxiety for end-users and not recognizing these anxieties could lead to implementation problems. Additionally, ERP systems are designed to replace legacy systems. The users of these legacy systems typically have vested interests, valuable experience and know-how in those systems. Replacing legacy systems means that people have to relearn new skills and their reluctance to do so might lead them to perceive the ERP system as being difficult to use. In other words, the more familiar an individual is with the legacy system the less likely the person will perceive the ERP system as being easy to use.

The main benefits of ERP systems are seen as the production of real-time data shared across the organization and consequently the integration and automation of business processes. This is particularly important in this new business environment where automation, effectiveness and efficiency in operations and real time data are important factors for business success. A link between ERP system modules operated and the perceived benefits achieved relates to the underlying reasons for adopting such systems. Increased demand for real time information, information generation for decision making and integration of applications are reasons for adopting ERP systems indicated by the majority of the respondents. With an integrated ERP platform in place, a business could build its enterprise applications on top of it. These applications could provide a timely feedback to enable optimal responses to changing conditions of customer demand and manufacturing capacity. ERP adoption times typically take from a few months for firms accepting all default settings, to years for firms needing to make major modifications. Because ERP has not made it easy to integrate other competing, best in class applications, most firms either face the high cost of modifying the ERP modules to meet the requirements or not install the applications. ERP systems thus with a lot of specifications has been implemented in the IT backbone. It is very important to bring compatibility between task and technology in the ERP system.

The ERP solutions seek to streamline and integrate operation processes and information flows in the company to synergize the resources of an organization. The disparate transaction systems are replaced by a single, integrated system that embodies the tight interdependencies among a firm's functional units. The majority of large worldwide firms have already adopted the solution of ERP. SMEs are considering it as a cost-effective and a competitive necessity to follow suit. Business processes such as logistics can become the basis from which to create a natural interaction of functional areas and business processes. The intersection is easily identified. ERP systems become the linkage within the organization that covers all major business processes, touching all the main functional areas in a business school curriculum. ERP adoption has three categories as: Infrastructure, capability which consists of improvement of processes, data visibility and performance which consists of cost reduction, strategic decision making, and adaptability to client requirements. The adoption of an ERP system can be the result of pressure exerted on the enterprise by its environment. A business that operates in a market very sensitive to price variations cannot afford high margins, and it depends on its IS for tight control of its production costs; it is a matter of cost transparency, and integrated systems are meant to satisfy this preoccupation. An enterprise operating in a strong growth market is made to reconsider its business processes in order to deal with its rapidly increasing size; the same is true of a business operating in a very dynamic sector, such as in high technology: the need to react rapidly to change accentuates the need for integration. That is similar to the environmental uncertainty construct, measured in its three components, heterogeneity, dynamism, and hostility. The need to optimize the supply chain is one of the factors that lead to the adoption of an ERP system: large manufacturing companies exert pressure on their suppliers, mainly SMEs, so that they will meet world standards in terms of cost, efficiency, and quality. This very often obliges these SMEs to restructure their processes, and in so doing them generally need an ERP system capable of real-time sharing of detailed information with their partners in the value chain. It is in fact the existence of very close logistical links between an SME and its business partners that creates an urgent need for integration. It is perhaps the absence of these links in the SMEs that may explain why these firms considered extra-organizational links with clients and suppliers less relevant in the process of selecting an ERP system, even though extended enterprise systems are specifically meant to integrate these links. The introduction of ERP has led to outcomes that are the opposite of the intended increase in managerial control. They suggest that ERP in the context of a global enterprise can be characterized as a runaway device, which behaves in unforeseen, erratic ways. Although they can be counter-productive, these erratic effects are not necessarily negative. For instance, the re-engineering of business processes within the oil company appeared to be self-reinforcing integration generating more integration as a side effect of the ERP implementation. In this case ERP had similar unexpected consequences for other IT systems, like the hardware infrastructure and the use of Lotus Notes software. In particular, the introduction of Lotus Notes significantly improved the necessary collaboration throughout the company, which also affected the corporate culture. Thus ERP is an important area to deal with and is a part of SAP and helps in creating a major focus in developing any internal organizational prospects.

CHAPTER 7

Supply chains of ICT and manufacturing firms

(BY BAISHAM CHATTERJEE)

To understand the main idea behind supply chains it is first important to understand the value chain and its effects and the results a very effective value chain can produce. In the modern world modularity has taken the pace of change. The popular and business presses have made much of the awesome power of computer technology. The modularity of the computer is in its increase of processing speed and storage capacity that helps companies manage this increasingly complex technology and thus make it easier to manage the assets of a business. Modular systems are much more difficult to design than comparable interconnected systems. The designers of the modular systems must know a great deal about the inner workings of the overall product or processes in order to develop the visible design rules necessary to make the modules function as a whole. The function of modularity can be better understood through the technology integration where the technology or the product can be divided or subdivided on the basis of: complexity, time for response in the market, customer preference, competitiveness, speed and ability to penetrate the market or situation and environment. Complexity depends on all this and more the complexity more would be the necessity to break even the qualities of the product to make it more effective. This situation has developed further and in the modern world and it can be better understood by breakthroughs in material sciences and other fields that have made it easier to obtain the deep product knowledge necessary to specify the design rules. The cost of computing can be better understood by the cost of storing, capturing and processing the knowledge and reducing the cost of designing and testing different modules. Modularity boosts the rate of innovation thus increasing the uncertainty which can be better assessed by Porter's uncertainty analysis. This is an important statement to quote about the uncertainty. Moreover leaders must also redesign their internal organization. In order to do this and to create superior modules, they need the flexibility to move quickly to market and make use of rapidly changing technologies, but they must also ensure that the modules conform to the architecture. Modularity can also be managed by team division and undergoing training and development during the breakup of teams into such groups as core design teams, support teams etc. Through careful directions the objectives can be defined by understanding the reason of failures in communications in the HR and leadership perspective in the value chain, as well as understanding the retentive ability of a business by understanding cost effectiveness and how other businesses perform in dynamic markets.

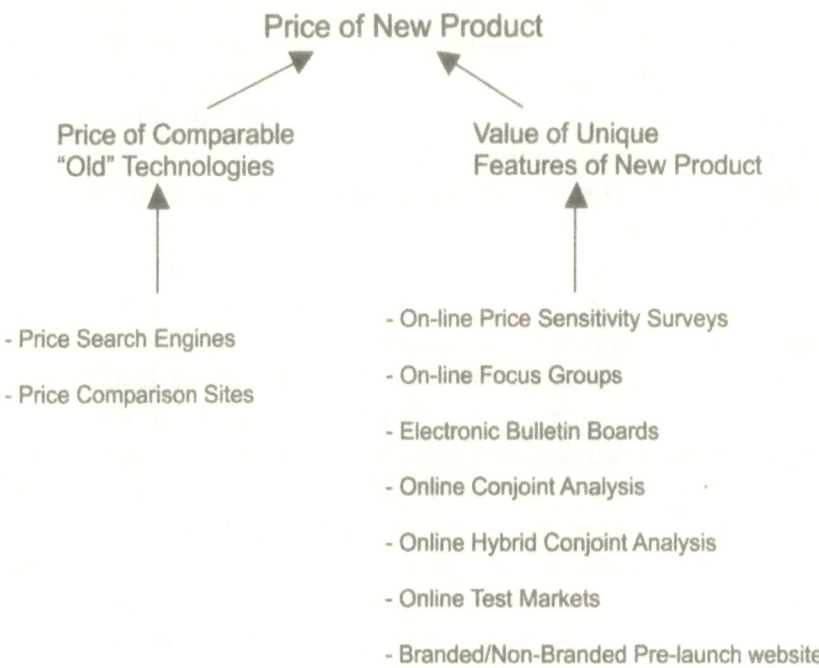

The Li&Fung case is a very creative outlook on the effectiveness of value chains, there are many others too. The supply chain in the case of Chrysler is also something to look at. Suppliers are much more organized to product development and in the drive for continual improvements of production processes. To get access to quicker activities and speedier supply chain processes manufacturers need to pick products in a competitive bidding and process. There are many suppliers involved competing for a big business and to understand the competition, the manager should effectively manage the bargaining power and its dependant rules as in the Porters 5 forces, as well as understand the bargaining power of buyers and the relationship between the two. Another way is pre-sourcing suppliers earlier in the development stage and these suppliers are responsible for building prototypes during development and for manufacturing the component or system in volume once the product is in commercial production. It is said that depending on the process characteristics it can be found that if suppliers are chosen then single supplier can be accountable for design, prototype and production parts. If a single supplier is assigned a particular task then there seems to be substantial investments in coordination mechanisms and dedicated assets. If this is possible then long-term contracts evolve and the necessity to make performance evaluations increase in order to keep the necessity of a supplier important and the chosen suppliers performance is very long term. Recognition of suppliers needs to make a fair profit is important in terms of relational characteristics and thus help in making considerable performance expectations and look forward towards cooperation.

In the modern world, never has so much technology and brainpower been applied to improving supply chain performance. The point-of-sale scanners helps the companies capture the customer's voice and electronic data interchange lets all stages of the supply chain hear that voice and react to it by using flexible manufacturing, automated warehousing and rapid logistics. Accurate response, mass customization and agile manufacturing offer models for applying the new technology.

It has been said that supply chains have failed for a long time trying various situations. But around 10 years of effective research has helped to devise an effective supply chain. Within this many ideas are important like—product life cycle (which varies and are difficult to be understood when prototypes are developed), demand predictability (economic factors and economic paradigms regulate this), product variety and market standards for lead times and service are the other key issues to be dealt with. Although innovation can enable a company to achieve higher profit margins, the very newness of innovative products makes demand for them unpredictable and increase the short life cycles and thus increases unpredictability. Demand predictability depends on other factors like market mediation costs that arise when supply exceeds demand and a product has to be marked down and sold at a loss or when supply falls short of demand, resulting in lost sales opportunities.

Traditional views of the value chain dealt with the models of the industrial economy but the modern thinking deals with ideas like global competition, changing markets and new technologies. New technologies can create a totally different mode of thinking in the technology segment, whereas global competition and changing markets have a much broader influence of creating a much broader and more organized marketing that will totally dominate the logistics. Ideas like global satellite and internet marketing can be added to it to create more value to the sales as well as deriving quicker customer satisfaction. This is another phenomenon of reinventing value and E-commerce strategies are the best ways by which the main philosophy of creating market segments and quicker relations can be made better. The tangible and more predictable way of creating value is by work-sharing, co-productive arrangements the company offers to customers and suppliers alike force both to think about value in a new way-one in which customers are also suppliers (time, information, labor, transportation), suppliers are also customers (business and technical services) that together creates more value per person. This is a whole idea of how value should be rearranged and the contributions necessary in the bargaining power to buyers and suppliers that is a key area of study in the value chain.

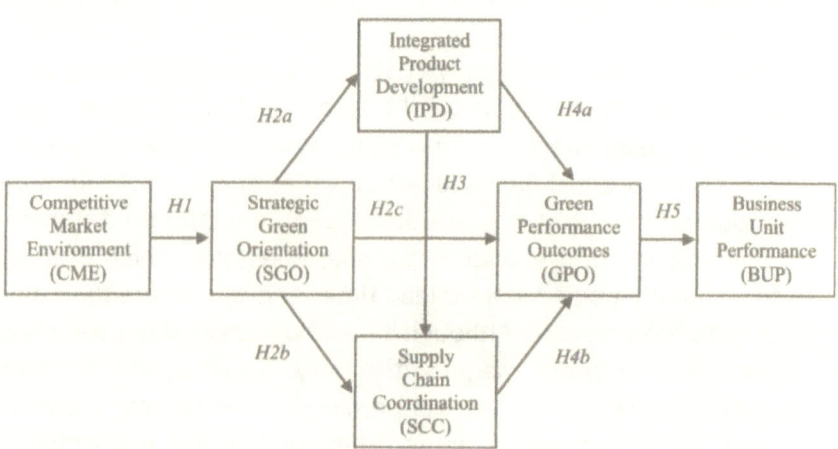

There is a lot of analysis on supply chains of manufacturing firms which started from experiences of original equipment manufacturers (OEMs) and their larger component suppliers but less is known about what happens in smaller companies. Japanese firms have used the idea of improving efficiency, reducing stocks and work-in-progress and then focus themselves on their suppliers. In order to remain competitive the OEM with all its resources nevertheless needs to concentrate and become yet more specialized. Similarly environmental regulations also address the supply chain by fulfilling socially responsible business practices. Producing environmentally sound products (ESP) and servicing them to the final customers demand effective coordination through the supply chain. To look at the knowledge perspective, green supply chain management which conceptualizes commitment on environment initiatives enhances firms' environmental performance through inter-organizational collaboration with business partners and increase efficiency by cost saving programs and proactive risk management practices. Strategic green orientation (SGO) is also innovation driven and requires inter-organizational innovation initiatives that intend to ensure complex performance outcomes through effective product design, supply chain integration and business processes. (SGO) is an organization's long term commitment for producing environmentally sound products (ESP) and services through the implementation of environment goals and programs in past, present and future. What also matters is integration of product and process design and prototyping and testing at early stage of product design. Supplier relationship and collaboration with suppliers and customers to bring a aggregate framework and calculation through statistics can bring a strong relationship.

Supply chain performance and effective management of supply chains have been increasingly recognized as critical factors in gaining competitive advantage of firms. Its measurements include costs, customer responsiveness and activity time. Most of these studies had mainly relied on the use of cost as a primary measure of supply chain performance since it was easier to implement in quantitative models. Supply chain performance has three types of performance measurement system like flexibility, resource, and output. Supply chains respond to change in products, delivery times, volume, and mix. Supporting this view, the findings of the current study show that supply chain design has a higher influence on the supply chain performance measures compared to integration and information sharing. To obtain the desired performance from a supply chain, the number of suppliers and their capacities, distribution channels, and the entire chain should be suitably arranged for meeting the current and potential needs of the customers, and the costs along the supply chain (inventory holding, transporting, operating, etc.) should be minimized. A well-designed supply chain in terms of locations, distances, capacities and planning can provide competitive advantage for the firms in that particular chain.

Few of the critical ideas in supply chain management deals with by saying in most businesses the supply chain management doesn't have a clear idea as to how the business should make use of procurement, marketing and what quantity of raw materials the manufacturing should make use of. Manufacturing has no idea what sales is seeing in the marketplace and procurement has no visibility into what marketing plans to promote. There are various tools that are always used in SCM. First of all the SCM tool used is supply planning tools, which help to align all the resources and activities required to get goods to market cost effectively. Secondly it is the demand planning tools that helps us in understanding through sophisticated modeling about the market demand. Plant scheduling tools helps in production plans whereas logistics systems help in the modern order management and inventory transportation.

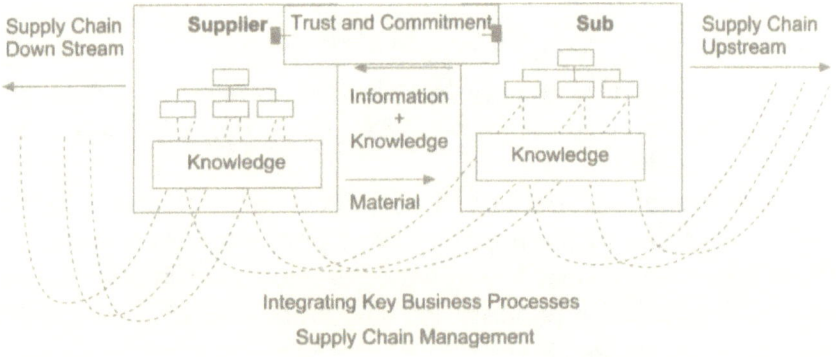

Integrating Key Business Processes
Supply Chain Management

Supply chain models identify three types of performance measures like resource measures, output measures and flexibility. Customer satisfaction, return of trading assets and flexibility are the measures of SCM performance. Agility refers to rapid strategic and operational changes to large and unpredictable shifts in the business environment. Flexibility depends on product, volume, launch, access and responsiveness. Moreover the SCM complex issues and the level of SC integration, competition capability also incorporates the firms performance. Internal SC integration is possible by automation and standardization of each internal logistics function, the introduction to new technology and continuous performance control. External integration is characterized by strategic linkage with suppliers and customers and standardization of logistics process between firms. A very important study shows that: sales, market share, and market position are influenced by not only advertising, competition level, product pricing and positioning, and degree of innovation in product lines, but also purchasing factors, thus emphasizing purchasers' strategic impact on the firm.

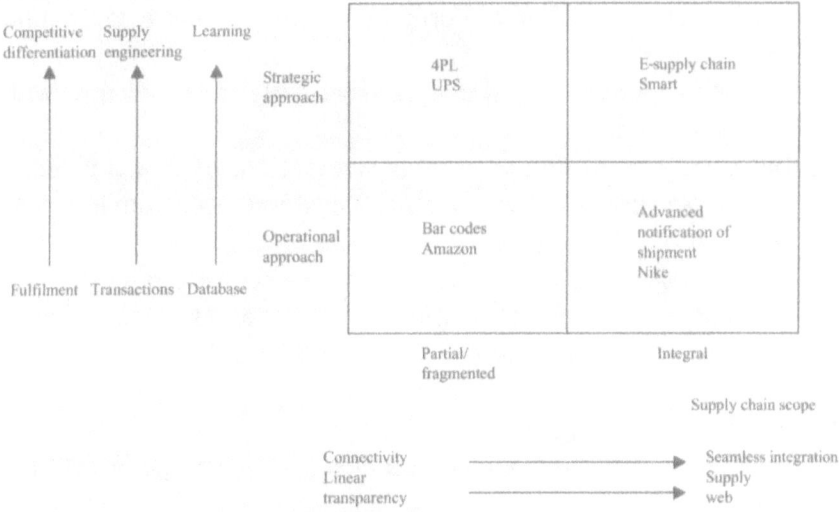

In order to measure all this, it can be found that supply chain relationships and manufacturing performance are more mixed. Transaction specific investments lead to an improvement in product quality and a reduction in product cost, but have no effect on flexibility or delivery performance. If managing the dynamics of a relationship are focused on achieving very high conformance to specifications, it can be more difficult to achieve performance on flexibility dimensions such as rapidly accommodating a sudden increase in demand (volume flexibility) or frequent or rapid changes in the product mix (variety flexibility). But the key drivers that put a business forward by concentrating more on the supply chain are:

(1) More efficient inventory control and improved delivery service;
(2) Improved co-ordination of material flows;
(3) Enable efficient re-planning and handling of product returns;
(4) Improved traceability;

(5) External requirements from customers and authorities; and
(6)Reduced counterfeiting or theft.

There are many ways of SCM development, in manufacturing firms, which are as:

(1) The internal supply chain integrates business functions involved in the flow of material and information from inbound to outbound ends of the business.
(2) The management of dyadic or two-party relationships with immediate suppliers.
(3) The management of a chain of businesses including a supplier, a supplier's supplier, a customer and a customer's customer and so on.
(4) The management of a network of interconnected businesses involved in the ultimate provision of product and service packages required by end customers.

These are all an aggregate details of how to get along in the SCM of a manufacturing firm. There are also other suggestive details for SCM in ICT firms:

(1) The application of postponement throughout the entire chain (in purchasing, manufacturing and shipment);
(2) The integration of the chain via information flow; and
(3) The extensive degree of outsourcing and subcontracting

The other virtual integration and postponement ideas that carry a firm forward with the different factors and conditions of postponement are:

(1) Time postponement (involving the delaying of activities until orders are received in time);
(2) Place postponement (involving the delaying of moving goods downstream in the chain until orders are received, thus keeping goods centrally and not making them place specific); and
(3) Form postponement (involving the delaying of activities that determine the form and function of products until orders are received).

Probably there are many modules for supply chain like inputs/ modules/products in progress that would be completely related to suppliers and distributors. Purchasing consists of many important points like involvement of suppliers engineering and operations, order-delivery lead time and expanding supplier involvement up to international investments. Manufacturing has also got to look at involvement in customization, consolidation of suppliers and look at time modules in progress and project management. Other things involved are postponed configuration, seamless integration with manufacturing and customer service, assuring lead time and customized response and consolidation of individual shipments. It is said that sales, relationship management and information search provide the key objective to new market knowledge that helps in growing performance in international markets and logistics is a growing idea of interest in the modern world and GIS, EDI, computer modeling dominates this growth.

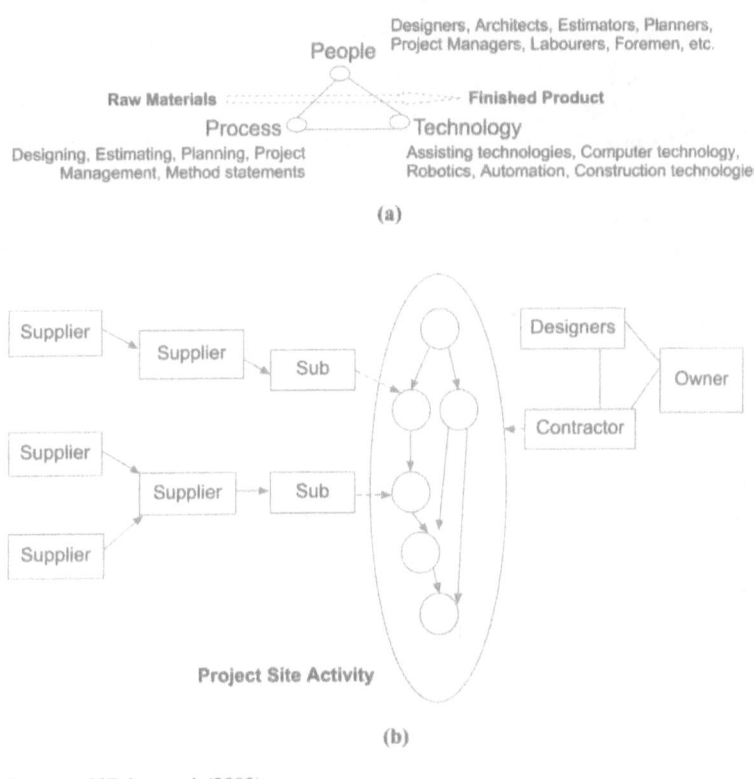

Source: O'Brien *et al.* (2002)

About e-supply chains that can measure and create the quickest benefits of modern times: The e-supply chain is the physical dimension of e-business with the role of achieving base level operational performance in the physical sphere (fulfillment, etc.). Additionally, it provides a backbone to help to realize more advanced e-business applications that companies will obviously be unable to achieve if base level performance is not even up to market requirements. Some of the e-supply chain failures are because: Supply chain failures are compounded by a lack of rapid reporting structures to identify emerging stock shortfalls, order processing systems to manage the needed repeats, and vendor quality management programs to ensure minimal complaints and returns. These companies have dressed the shop window with slick Web sites, but there is no technology behind it to fulfill the order. Systems are already inadequate for big volumes, before adding new developments using mobile phones to access the Web. Companies are creating a demand they simply cannot fulfill.

Nike has developed a supply-chain-wide advanced notification system for its products along with European distribution centers. They are developing closed sites for its long term preferred suppliers in an attempt to leverage transactional data and open databases relevant to order fulfillment. There are many supply chain redesigning processes that makes the work easier.

The important points stand as reallocating the roles actors perform in the chain and related processes, eliminate activities that do not add value, redesign policies and improve the supply capacity and exchange all the information like inventory, work in progress etc. It is important to increase the number of events per unit time for all processes, establish an information exchange infrastructure. Other very important positions of redesign are change position of chain decoupling point and increase manufacturing flexibility. Other ideas involved are differentiating to products, systems and processes and simplify structures. Other ICT objectives to redesign are: implementing real-time ICT systems for information exclusion, standardize bar-code, jointly define logistical chain objectives and chain performance and look at incentives and performance along with chain objectives. In another important idea behind redesigning it has been found that transformed organizations, that open system information for customers, suppliers and partners, it has been found that e-business that integrates and minimizes waste at every step in the supply chain evolve from e-commerce that is an evolving creative technology for modern redesigning and e-mails, communications and websites are the main supportive ideas to this.

Another important area of study is the supply-chain co-coordinator that can provide the necessary means to manage and control the supply chain. They help in certain ideas starting from marketing basics 5Ps, cost of distribution, information, assessing and settling conflicting interests of sellers and buyers and understanding characteristics like product matching. Electronic service centre (ESC) is something that helps in performance of SCC and it integrates existing information system with own information system. In this process messages submitted between organizations include purchase/sales/transport orders, order confirmation, catalogue updates, receipt orders, pre-alert notices, pick-up orders, status updates, exception reports, confirmations of order fulfillment, proof of delivery (POD), receipt information, order closed, pre-invoice summaries, invoices and statistical information. A pre-alert notice can be used so that organizations can make the necessary arrangements to receive goods or to make goods available for transport. Status updates, exception reports and POD messages are used for tracking and tracing. "Tracking" refers to getting up-to-the-minute information on the whereabouts of a certain load, and "tracing" refers to the evaluation of the routes followed over a period. Financial information is consolidated in periodical (often monthly) pre-invoice messages. Statistical information includes the number of orders, types of order, and quality of service. External uncertainty is based on control systems and manages a relationship in information flow and information gathering and look forward towards flexibility and risk.

There are many methods of understanding flexibility dimensions and one of them is product flexibility, which is a very frequent literature. The second type of flexibility touted in the operations literature is volume flexibility, defined as the ability to effectively increase or decrease aggregate production in response to customer demand. It requires close coordination between a manufacturer and its suppliers, especially in the face of increasing demand. Volume flexibility directly impacts supply chain's performance by preventing out-of-stock conditions for products that are suddenly in high demand or by preventing high inventory levels. Delivery flexibility is another important point where the company's capability to adapt lead times to the customer requirements. Trans-shipment flexibility involves movement of stock between locations at the same echelon level. Postponement flexibility implies the capability of keeping products in the generic dimension. Sourcing flexibility is finding another supplier for raw materials. Sometimes it is true that responsibility is spread throughout the supply chain and effective performance hinges at the firms ability to leverage the capabilities. Another flexibility is launch flexibility that helps in introducing new products and product varieties. There is something known as the supply chain planning that requires stronger relationships to allow the collaboration required on non-specific product groups. SCP and redesigning helps in building close relationship that facilitate services geared towards the manufacture of bespoke products and high level of collaboration between partners creates customized commodities. Very often it is felt that the ICT infrastructure and investments support hardware support that leads to reliability, interoperability and EDI that helps in the efficiency gains of the most talked about things in modern world like ERP, CRM, MRPII and OLTP that affects profits.

E-supply chains are becoming the most talked about factor in today's world where it is necessary to have the ability to adapt and engage in both short-term arms length and long term collaborative planning. Standards depend on nature of agreement, planning depends upon the needs of the focal organization. Responsibilities relate to inter firm exchange and specification of the goal. E-supply chains have long term partnerships in comparison to traditional SC organizations and internet based e-supply chain. Thus purchasing, procurement and business logistics create the foremost impact on developing the correct ideas towards redesigning, planning, integrating the supply chain.

CHAPTER 8

Interrelationships and economies of scale utilizing Porters competitive advantage
(BY BAISHAM CHATTERJEE)

Interrelationships have long been present in many diversified firms. Achieving interrelationships involves far more than simply recognizing their presence. There are a number of organizational barriers for achieving interrelationships in practice, which are difficult to surmount even if the strategic benefits are clear. There are 3 types of interrelationships among business units: tangible, intangible and competitor. Tangible interrelationships arise from opportunities to share activities in the value chain among related business units, due to the presence of common buyers, channels, technologies and other factors. It leads to competitive advantage by sharing lower costs or enhancing differentiation enough to exceed the cost of sharing. Business units that can share a sales force may be able to lower selling cost or provide the salesperson with a unique package to offer to the buyer. Thus this type of interrelationship involves sharing multiple activities and cross-sell each other's product through the medium of joint sales force. The second part that is the intangible interrelationships lead to competitive advantage through transference of generic skills like type of buyer, type of purchase by the buyer, type of manufacturing process employed and type of relationship with government or know how about how to manage a particular type of activity from one business unit to another. Competitor interrelationships stems from the existence of rivals that actually or potentially compete with a firm in more than one industry. It helps in making the other two types of interrelationships more important to recognize and exploit. Sharing and differentiation and the advantages of sharing and business unit position are very important. A cost advantage results from sharing fabrication of the component with a business unit that uses a small volume of it. The costs of sharing consist of costs of coordination, cost of compromise and cost of inflexibility. The cost of compromise is frequently reduced if an activity is designed for sharing rather than if previously separate activities are previously combined or if an activity designed to serve one business unit simply takes on another with no change in procedures or technology. Inflexibility consists of competitive moves and exit barriers. Another idea is known as procurement interrelationships that involve the shared procurement of common purchased inputs. Technological or production interrelationships have their impact on cost or uniqueness of technology development, sharing product or process technology through the value chain can also be identified. The last type of this interrelationship is known as infrastructure interrelationship that involve firm infrastructure and involve such activities as financing, legal, accounting and human resource management.

Controlled decentralization encourages business unit managers to pursue strategies that maximize unit performance and not the corporations and undermines the pursuit of interrelationships. Tangible interrelationships require sharing activities in the value chain to achieve this goal and intangible interrelationship requires transfer of know-how among business units for the pursuit of this goal. A balance must be struck between the vertical and the horizontal elements in a diversified firm if the potential of interrelationships is to be unleashed. The number of interrelationships is of different types like horizontal structure that cut across business unit lines. Horizontal systems or management systems that have a cross unit dimension in areas such as planning, control, incentives and capital budgeting, horizontal human resource practice that facilitate business unit co-operation undertakes a horizontal conflict resolution process. There are many impediments that stop from achieving interrelationships. Highly centralized firms with many small business units, firms with a strong tradition of autonomy, firms built through the acquisition of independent companies, provide no effort to create corporate identity, which earlier had a bad experience in pursuing interrelationship. There are also few other ways of achieving interrelationships like greater tradition of committees and frequent personal contact among executives, intensive and continuing in-house training and corporate wide hiring and training. Thus taking the industrial perspective a very good example of interrelationship between manufacturing firms and their accommodation is increasing the size of the letting market, providing sufficient comparables for assessing rents at review and minimizing physical obsolescence and management problems. Contemporaneously, on the supply side, changing construction techniques have allowed the development of adaptable, inexpensive, single-storey, light industrial buildings. Shorter construction periods permit more flexible adjustment between supply and demand. Overall depreciation rates have been reduced as lower site coverage has increased the land component of total investment value. Finally, industrial investments have a lower unit capital value compared with other sectors of the property market, allowing a spread of investment and minimizing risk. All these factors allowed industrial property to establish itself as an investment acceptable to institutions and other investors. There are three main factors which determine the demands made by the firm on its building. First, intensity of production plant, defined as the density of production equipment in the area exclusively used for manufacture and associated materials handling. A high figure implies a machine-intensive production system such as that found in bulk process manufacture or high speed production line manufacture e.g., bottling. Conversely, plastics extrusion and clothing production are examples of industries with low densities of production equipment. The second factor is the rate of change of production plant. The frequency of significant alterations to production machinery and services, and to fixed path mechanical handling equipment, affects the level of adaptability demanded of the building structure. High rates of change are those where change occurs every year or more frequently; changes which occur every 20 years or over a longer period are considered to be low rates of change. The third factor is the intensity of use of mechanical handling equipment in production areas. This is a guide to the demands made upon his/her accommodation by the manufacturer. Bulk process pipes/hoppers are examples of a high intensity use, while a single monorail hoist indicates very simple handling requirements.

TMT is usually composed of key managers who are responsible for the making, planning and execution of business strategies and consists of different management styles which consist of business operations like organizing, planning, leadership, staffing, stimulation, control, communication and authority and delegation. TMTs are major decision makers of corporate strategies and operational abilities with direct impact. Issues about cooperation, innovation, and organizational learning have been investigated from a variety of dimensions and perspectives with lack of dimension in organizational operation. Ambitions and innovation undertake a risky venture contrary to a conservative posture. In the era of knowledge competition, if firms do not absorb knowledge quickly or manage and process knowledge effectively they will not be able to escape the fate of being eliminated from competition. To understand the interrelationships in case of performance measurement systems, scholars have recommended that greater clarity in the linkages between different dimensions of the organizational performance should become an important issue during the process of designing performance measurement systems. As organizations increasingly use causal models like strategic maps for the basis of their performance measurement systems, methodologies already established to develop causal models can be more widely employed to help in the design phase of performance measurement systems.

The process of building a corporate data model consists of: requirements analysis, conceptual framework, logical model, system implementation and testing and revision. This is a method of building interrelationship. The diagrams are intuitive and easy to follow, and the attribute table provides the details needed for closer inspection. It provides a common language for users at different function levels; it also allows the transfer of technical knowledge to the general public. While the visualization approach is an effective management tool in organizing business rules; the clearly stipulated sequence of the business rules can simplify the task of constructing a business rules model. When the reassessment of business rules becomes necessary, the ER diagram and the production rules can help the database administrator to see clearly all the business rules that require maintenance. Documentation is a very important part of business rules modeling; it can be used either as a user's manual or as a guide for troubleshooting.

Another area of study is the interrelationships between network externalities and information asymmetry as well as between e-business adoption and information asymmetry. It is proposed that network externalities would significantly influence e-business adoption, and that e-business adoption and network externalities would reduce information asymmetry. It is also proposed that cultural contexts would have a significant influence on the interrelationships among network externalities, e-business adoption, and information asymmetry. Several models and theories have investigated factors that affect e-business adoption, such as innovation diffusion theory, institutional theory, the theory of planning behavior, the technology-organization-environment framework, the technology acceptance model, and the perceived e-readiness model. Factors investigated in these models and theories may be categorized into four domains: innovation attribution e.g. relative advantage, complexity, and compatibility, management characteristics e.g. top management beliefs and risk propensity, organizational characteristics e.g. centralization, formalization, specialization, and technological readiness, and environmental contexts e.g. infrastructure maturity and competition intensity. A distinction between information sharing and information collection is defined as follows: Information sharing refers to the sharing of information among employees and with supply chain members, while information collection refers to the collection of information from organizations outside of the firm's supply chain. The distinction between sharing and collection is that sharing implies mutuality, while collection is only one method of gathering information outside an organization. In today's competitive environment of global and digital economies, supply chains, rather than individual companies, compete with each other. The growing size of e-business adoption facilitates information exchanges among vertical supply chain members and horizontal peers and entities outside the supply chain. Accelerated information flow and information exchanges allow companies to be more knowledgeable about their products, pricing, and markets.

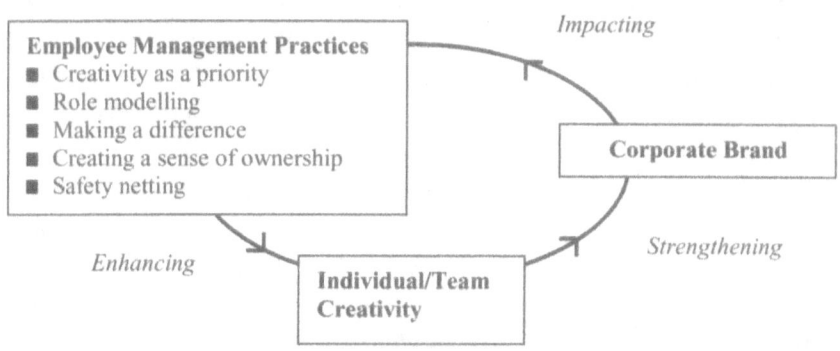

Moreover, the relationship between organizational members that brings corporate identity and a reputed brand consists of various leadership activities, flexibility, feedback, sufficient resources and financial support all as signs of creativity. Three major components and their interrelationships are explored in this framework, namely employee management practices, their influence on individual/team creativity and corporate brand building. Moreover while talking about entities; the definition of an entity is a concrete abstract thing of interest. It includes associations among things and is perfectly suited to the encapsulation of ideas and concepts. It allows that animate and inanimate objects as well as abstract notions are rendered equivalent and it therefore liberates us from the usual constraints of thinking only about one type of thing at a time. The relationships between entities allow an indirect reference to function, without being explicit. This allows us to liberate ourselves from simple process models which mask the underlying concepts. Process models are natural and comfortable for people to work with, but they tend to reinforce current practice and obscure the more enduring ideas behind a business. A strategy- target market type of idea is also a type of interrelationship where each strategy combination either does or does not make sense conceptually when real-world considerations are weighed. For example, the control and centralization that is required for a low-cost approach supports a one-size-fits-all standardization strategy and is appropriate for a target market of international sophisticates. Whereas low-cost production and standardization targeted for acceptance in a provincial customer market are contradictory. The global strategic marketing decision tree is compelling from a strictly practical position. Examples of leading multinationals combining the six conceptually compatible combinations are abundant, while few or no cases can be found in support of the incongruent combinations. Studies on market-orientation have an operational focus on market information process activities, which is mainly on the subject of customers and competitors, particularly information acquisition, information distribution, and the ability to behaviorally respond to what has received. They further indicated that market orientation could be characterized by the degree to which firms acquire, disseminate, and respond to information gleaned from customers, channels, and competitors. Beyond the concept of learning orientation, the concept of organizational learning and learning organization has flourished and been defined in a wide range of literature. The five key factors to promote organizational learning: system thinking, personal mastery, mental models, shared vision, and team learning argued that as the world becomes more interconnected and business becomes more complex and dynamic, work must become more learning. The organization that truly excels in the future will be the organizations that discover how to tap people's commitment and capacity to realize their highest aspirations argued that traditional management strategies are oriented toward building advantages over the competition while new management strategies in a dynamic changing environment need to stand on a more revolutionary viewpoint and make breakthroughs and quantum leaps to gain a leadership position. Based on this viewpoint, it is suggested that the paradigm of organizational learning needs to shift from single-loop or double-loop learning to triple loop learning or unlearning, from knowledge creation through incremental changes to knowledge creation through radical changes, from system thinking to creative thinking, and from continuous improvement to creative and innovative improvement. Thus, it is one of the purposes of this study to evaluate the paradigm shift of organizational learning and its influences on innovation and business performances. With the dynamic changes of management environment, the capacity of innovativeness has been regarded as one of the most important factors that impact on business performance. It is generally recommended that firms engaged in innovation will result in higher performance on product development, process renovation, and flexibility and responsiveness. The positional advantage may be decided by a firm's market orientation, learning orientation, entrepreneurial orientation, and innovativeness. However, relatively little studies have concentrated on the linkages among these constructs. For the relationship between market orientation and innovativeness, a market orientation essentially involves doing something new or different in response to market conditions, it may be viewed as a form of innovative behavior.

Innovation and new product success are more likely to result from being market-driven. These authors view innovativeness as the medium for business success in the wake of continued intelligence gathering and decision-making, after finding performance linked to both market orientation and innovation, speculated on a causal relationship among these research constructs. It shows that proactive information search will result in an organization's innovativeness. Innovativeness is one of the core value-creating capabilities that drive the market orientation-performance relationship. That shows that market orientation contributes to new product success. For the relationship between learning orientation and innovativeness, an excellent learning environment in an organization will leverage the use of all resources, including the activities that accompany a market orientation and innovation. The continuous orientation toward organizational learning will improve the efficiency and effectiveness of a firm's innovative activities. Firms must ensure employees to continuously absorb new knowledge, and maintain a superior internal knowledge management system, since knowledge is the key that combines organizational learning and innovative activities. For the relationship between learning orientation and market orientation, learning orientation refers to an organization's capability to contest old assumptions about the market, while market orientation refers to an organization's focus on environmental events that may affect their ability to maximally satisfy customers. The major distinction between these two concepts is that learning orientation does not just utilize market-based knowledge to promote customer satisfaction. The learning orientation can facilitate a firm to react effectively to external changes, such as customer preferences, and product technology. The enhancement of learning capability will enable the organization to absorb and assimilate new ideas that viewed that learning orientation is a precursor to establishing a culture that is receptive to innovation.

Economies of scale arise from the ability to perform activities differently and more efficiently in larger volumes, or from the ability to amortize the cost of intangibles like advertising and R&D over a greater sales volume. Economies of scale reflect not only the technology in a value activity but also the manner in which the firm chooses to operate it. In the case of a manufacturing business despite the indubitable importance of flexibility, it is argued here that economies of scale remains an essential determinant for cost-efficient production, and that, without sufficient EOS, high levels of flexibility cannot alone translate into world competitive production. It is assumed that EOS is only optimal when a plant is operating at high capacity utilization levels, and that where this is not the case penalties are severe. As a rule, EOS arise from the fact that an increase in output of one or more products does not give rise to a proportional increase in cost hence the longer the production run over a period of time, the lower the unit costs will be, until the minimum efficient scale of production for the products in question has been reached. Most economies are internal to the manufacturer, although external economies also exist. Internal economies also known as technical economies relate to production costs and emanate from a variety of sources. Economies arise from the fact that, generally speaking, costs increase with surface area, while output increases with volume or capacity of additional units of plant and equipment—hence, for example, a 100 per cent increase in output will necessitate only a 60 per cent or 66 per cent, depending on the rule adopted increase in costs. Economies of specialization whereby higher output provides opportunities for greater division of labor into specialist tasks, together with increasingly more sophisticated equipment which further specialize hence quicken and improve quality of tasks. In economies of scope economies accrue from transfer of knowledge across different, but related, product lines. The principle remains the same for these as for EOS, i.e. the larger the manufacturer, the greater the ability and opportunities to achieve economies. Indeed, under the prevailing situation of rapid technological change, economies derived from R&D and from promotion have become increasingly important.

To give more details about the internal factors diminishing cost of the mix of factors needed to supply a single specific service or internal economies of scale, on to cost reductions on a bundle of related services or internal economies of scope. Benefits of increased market penetration, with economies of scope are used in developing a customer link thus potentially economizing on marketing and related costs. A higher level of economies of scale in production costs is expected to create a tendency towards opening few large facilities, each operating at a high level centralization, while under diseconomies of scale of linear production cost functions, we expect many facilities to be operating at lower levels decentralization. It should be noted that consideration of production economies of scale in solving transportation problems becomes most vital when dealing with location-allocation problems, where fixed charges of opening facilities magnify the effects of economies of scale. To recognize the underlying industrial structure economies of scale and scope and network externalities the industry level demand and supply either has a critical mass for focusing, or offers possibilities to realize economies of scale. In mature industries economies of scale can be achieved either with investments in manufacturing and vertical integration or by mergers and acquisitions. In emerging industries the dominant standard is provided either by regulators or by users can give the needed first mover's advantage to profit from the economies of scale and scope.

CHAPTER 9

Linkage: An advantage of quicker inter-networking that brings competitive advantage
(BY BAISHAM CHATTERJEE)

Linkages are relationships between the way one value activity is performed and the cost or performance of another. It is a phenomenon that makes innovation easier. An example is reducing scrap and simplifying manufacturing by using precut steel sheets for further processes. Linkages help in gaining competitive advantage in two ways: optimization and coordination. It is also possible to obtain cost advantage, differentiation, substitution and technology through a linkage. In the modern world when creativity is there in every small business in the western world what matters most is attitude testing through Likert scales which is the most modern phenomenon that businesses throughout the world think of dealing with. To look at the HR perspective, linkages have a subjective influence on communication skills, interpersonal skills and promotability.

It is even found that strategic planning has linkages involved with internal processes and external processes involved. Strategic planning consists of internal and external environment; the external environment orientation to the planning process has to be noted. As well as the internal operations and maintenance staff, which conducted the internal analysis of the firm, to determine the internal orientation of the planning process? The two results of linkage to strategic planning are divided into two components: attention to internal facets, and functional coverage and integration. There are various techniques that lead to better linkage success like developing activity based costing, developing financial models that can derive benefits and success in the sales and marketing and develop scenario techniques based on uncertainty principles that would give the business a leverage to prove superior to others in the economy. There are different benefits that derive value like a superior work environment, superior learning and development and a negotiable pay. All these ideas coming into view with the nature of work force that can be analyzed on basis of the performance management decisions and dependency of bonus and revenues based on dollar terms. This helps in controlling the employee work benefits through compensation and long term training programs. Performance management is a result of this and to put in more details we take up the term performance measurement.

Similarly market demand and new products are linked by customer requirements and the need for the product at that particular point of time in the particular segment. Similarly this phenomenon can get imbalanced due to need for promotions which can only be managed by previous performance in the logistics system and utility of time. But to manage all these linkages many price related activities are involved that have cost of ownership, understanding specific supplier costs, target costing and market monitoring. Vendor analysis is another measure of creating a solid linkage to the demand chain. The paradigm shifts to incorporating demand chain with strategic procurement, managing structure and culture of vendor relationship. Energizing internal and external organizational teams through flexible structures and responsive information systems in the organization balance the supply chain capability with customer demands. International purchasing, addressing issues related to variability in the international environment and look towards conceptual models or benefits related to the country specific. Environmental linkages lead to integration of communications and technology, innovation inside the business and parallel sourcing. Similarly logistics costs are a result of vision and goals of a business and strategic partnerships in the distribution channels which creates this linkage between the two that not only drives growth, but also derives the idea to create better efficiency between inbound and outbound logistics. Similarly long term supplier collaboration has created a linkage between behavioral patterns and supplier association with a result of integration and ability to compete in the common marketplace. There are many ways of deriving linkages where logistics managers focus on inventory, facility location and design and transportation matters while finance managers obtain the lowest cost of borrowing. The result of linkages drawn among logistics and transportation shows this. A differentiation between supplier cost structures and product competition helps or encourages suppliers to share cost data that requires that suppliers reveal their books to the buying organizations to help identify cost drivers and high cost areas, and then work together to reduce costs. Target costing is another idea that helps in the minimization of total logistics cost and capital carrying cost. This brings out the price market will bear and then backs out desired profit and becomes the overall allowable cost for the product or service.

Similarly again returning back to using the Likert scales, two possible mediating factors are feedback-satisfaction linkage, job challenge and job involvement. This tends to make the sales job challenging. Constructive criticism helps in understanding the challenge and thus performs. This attitude of salespeople reduces the negative effects of criticism through feedback. There is a procedure of linkage that estimates and compares a direct effects model and a partial mediation model. It studies the positive and negative feedback that creates two processes job challenge and job involvement which are also related to the direct effects model that considers direct phenomenon to ideas. The other linkage that is created between the negative feedback-job involvement relationships creates a path co-efficient. There may be linkage between three elements like positive feedback—job challenge—job satisfaction where this channel is used to serve to enrich the supervisor's feedback, shifting it from the domains of simple encouragement toward supervisor-initiated development.

Similarly the element that can create the greatest value to the linkage is the supply chain. Real-time information exchange with suppliers in the upstream and with customers in the downstream will create an opportunity where optimization can take place. Linkage reduces the adverse effects and contributes to enhance performance.

The linkage in SCM helps in lowering cost, shortening delivery time, providing appropriate feedback, maintaining low inventory levels and improving reliability. The impact of supply chain linkage on supply chain performance has been studied for sometime. But more of the linkage strategies are based on supply chain strategies to reach supply chain goals that are the key idea for decision makers. After deciding on the SCM linkage the other thing to deal with is the customer linkage that is concerned with planning, implementing and bringing out successful relationships that is more related to the logistics part and area of interest of the SCM. Customer linkage helps the SCM department to bring products at the right time, right place and right quantity with correct invoice. The most important thing about customer linkage is that it helps in managing demand and order scheduling. This helps in the performance reliability measure and together by combining the SCM and CRM, it creates the internal linkage. After going through the linkages that create cost containment and performance reliability, we think of linkages that would initiate strategic alliances by integrating suppliers into the managements decision making process in an effort to reduce product cycle time, improve delivery cycle time and reduce subsequent costs. It's difficult to execute this type of linkage and it needs a lot of careful monitoring.

Survey based empirical work both related to internal outcome and external outcome helps in developing theories in related discipline. Similarly in this chapter on linkages, if low cost and product differentiation are linked it gives rise to quality, flexibility and delivery. Similarly in the preceding condition if structural and infrastructural decisions are linked there is consistent support for chosen competitive priorities. Similarly manufacturing objectives and the costing principle are often linked to create maximum outcomes and measurements in costs by arousing three principles: variable cost, material cost and overhead costs. Similarly comparing the manufacturing ability where there is a linkage between procurement lead time, new product development cycle and equipment changeover time what arises is quality and defect rates which has to be calculated and measured.

Cost reduction deals with prices of products and services. Similarly after the cost reduction process is determined the linkage between production efficiency and adopting new technology creates re-engineering production processes and lower price in the market.

Businesses that prefer a cost reduction strategy must rigorously control and minimize expenses and strive for greater economies of scale. Similarly the linkage created between cost leadership—utilization produced low values but is definitely a great advantage to innovation.

Environmental technologies that help in product designing to reduce negative impacts in the natural environment as they respond to the linkage between environmental technologies and government regulations that bring out the importance of public pressure and customer demands. Allocation of investments and availability of resources to create plant level opportunities is another important form. Similarly competitive advantage can be created through linkage where investment in manufacturing technology invariably forces operations to change as the new technology is adopted, integrated and leveraged for competitive advantage in the areas of cost, quality or timeliness. Firms that invested more heavily in quality-related systems also avoided pollution control technologies, with a greater part of the investment on pollution prevention and management systems. Pollution control is an environmental idea on quality inspection; monitoring and rework and product redesign, preventive audits, which are all examples of quality management through the manufacturing system. Similarly it is said that to find out and elaborate more on environmental innovation that relationship between knowledge management and R&D has to be evaluated looking at the KM process, transforming information and technological advancements and all these evaluations are somewhat dependant on knowledge communications. There are many linkages arising from the third generation research initiative that consists of emphasis of partnerships, cost-benefit analysis arising after the linkage with marketing and balancing of priorities is formed. Moreover it is said that the recent models deal with linkage between customer needs and core competence that help us in modeling for the future, and developing the architecture and capability and create dominant ideas that would forever bring the difference and bring in optimistic ideas. Moreover intelligent algorithms, strategic maps and full scale web based innovation are definitely the symbol of the recent and most future oriented and deterministic advancements in the modern world. The R&D is a technology based knowledge map to take into consideration the use of knowledge portal that considers the latest knowledge and business processes of the firm where the extent of knowledge and knowledge conversion process is required whereas R&D more depends on information resources and new technology predicted.

Similarly the investment and TQM models create the linkage through production layout and pull production system. Similarly linkage is very important in the case of knowledge accumulation, organizational innovation and performance where the linkage helps in bringing business growth and development closely affecting the external environment. Firms must keep control of different variables within the environment like knowledge applications, professional levels and experience. Similarly it means that a global competitive business climate and business scale brings a very creative innovation phenomenon. There is definitely a relationship between this environment that has been talked of and organization culture. As well as there is relationship between knowledge accumulation capability and organizational innovation. The effect of the external environment on the correlation between knowledge accumulation and organizational innovation, external environment and organization knowledge accumulation will result in a mutual interaction to influence the organizational innovation. Knowledge accumulation and external environment people have to face give a better advantage to the knowledge they possess to any changes in external environment, giving both administrative innovation and technical innovation.

Similarly it is said that a linkage in manufacturing strategy is based on the relationship between strategy (that consists of product markets, positioning and competitive features), processes that is a combination of the most modern manufacturing practices and structure that consists of integrating individuals, groups and units created among them. There are many characteristics for adopting a linkage: the first of them is strategic adaptation that plans and implements manufacturing strategy and provides a linkage between business strategy and organizational commitment to strategy. Moreover the linkage created in production system is through JIT that is through a linkage between rationality and production operations. Whereas there is a linkage created in the less talked about operational system for quality between working floor and systemization of QC activities that creates quality committed behaviors. Similarly it has been viewed that manufacturing companies have four main objectives, namely, flexibility, reaction time, quality and return on assets. Performance metrics, best practices and benchmark data help to validate on this. This data is divided into eight groups: employees, planning and control, production and assembly, research and development, distribution, order handling, purchase and suppliers and market and clients.

Automatic data collection is defined as collecting data by the firm without human input or minimal effort. It is facilitated by scanning devices and barcodes. This internet study was undertaken to provide some procedural guidelines to improve the installation of these logistical linkage processes with the most justified theory. The less likely the EDI adoption and linkage between the two firms more is the incompatibility with the present physical system. The ability to bring together channel members makes each firm more committed to the system of distribution. The concept of areas of linkage relates well with the most current business thinking, by which services can be segregated and more value can be added to customers. Competences similarly help in transfiguring or re-engineering or process engineering of customer services. Raw materials assembling and thereby economies of scale can be measured more carefully. The entire organization can express a shared interest in adopting EDI. Representatives taking up this action should discuss and agree on the implementation strategies and time-scales before implementation takes place. Electronic customer response which is a branch of EDI utilizes data sharing and co-operation to find out a linkage that would help in reducing distribution costs. Vendor managed inventory and activity based costing is an important part of ECR.

The dimensions of market tracking and research and development can be calculated through the firms' financial performance and these management practices ultimately lead to firm profitability. International competition is also related to the linkages of firm profitability when firms try to compete globally. At this point of time, the process of introducing new products becomes more complex and the process of production to market must reach at a much faster pace. Similarly looking at the financial records market size was introduced as a potential industry factor. Market size is a potential important factor of a firm's performance on three firm factors: sales volume, market share and conglomerate status. In a well defined market with huge competitive strength sales volume is considered as an important performance indicator. Improved services may be a prerequisite of a positive market orientation performance linkage. Similarly service quality and business profitability is market oriented. Similarly in the case of marketing and sales that is linked to communications firms with stronger linkage paid attention to creating a common business language and design the marketing and sales of an organization structure that creates the main support for discussion.

The value process relates to bringing a linkage between 3 aspects customers, organization and external stakeholders where customers are dominated by customer markets and referral markets; the organization by internal markets and recruitment markets and external stakeholders by shareholders, others those who influence markets and supplier & alliance markets. There are determined linkages that create relationship and create a positive link between customer satisfaction and customer profitability. There is also a significant relationship between customer satisfaction and loyalty. There has been a growing relationship between the customer and the employee particularly in relation to service businesses. Also while talking about the other CRM perspectives like employee satisfaction and customer satisfaction the linkage relationships created respond to the customer service that employees deliver. Similarly financial measures, innovation, employee measures and customer measures can be linked to one another where the service profit chain model that deals with knowledge enhanced, market centered, causally related, change related and the potentiality nature is determined.

Linkage is a term that is derived more from the value chain where even if the value chain is reconfigured with time there always seems to be an existing relationship between customer order processing and financial reporting system. The product group in an organization can be identified from raw material input to final delivery to customer and the existing value adding capability can be determined. It is found that inside the value chain both logistics and manufacturing play a great role. In the logistics section ideas like logistics impact on customer and logistics customer service coupled together to give the quality to the logistics. Similarly a triple linkage between delivery performances cost of rapid and reliable delivery and expediting performance (i.e. that shows the length of time to deliver expedited items) couple to form the delivery perfection of logistics. Similarly as I have talked a lot about marketing and sales and earlier about logistics flexibility is formed through quick response that shows the length of time to respond to customer enquiries and cost of flexible and responsive logistics system. Similarly cost structure or cost information of a logistics is formed through a linkage between total cost information, transportation costs and order system costs that can calculate the cost structure of a single logistics. Similarly innovation is formed through a linkage between logistics cycle time analysis that reduces cycle times, value analysis and cost of logistics service innovation. Similarly manufacturing information is based on process control, defect rates, costing, total labor costs, backorder performance, cycle times and R&D effectiveness. Similarly a logistics or manufacturing cost structure, cost measures and performance is based on understanding the integration and coordination and forming this into a linkage with learning and experimentation and innovation and transformation. Similarly the logistic centre looks towards information flow that comes from customer knowledge, cash and product that has to be maintained through direct contact with customers, technology standards and developing service oriented product. Similarly it is found that through the success of one of the factors the other factors or conditions come into action in a business process reengineering principle. In the first category of system analysis and design: simulation, prototyping, spreadsheets are the key initiatives for success. Similarly in the operations research and work study methods: it is very often that researchers talk of ergonomics, time and motion studies and statistical process control. Information technology, change management and human resources are other key factors for success of linkage and creating competitive advantage. Organizations already employ a mix of performance measures and what needs to be explored is a broader spectrum which takes into account non-traditional, especially non-financial approaches to performance measurement and thus creating a linkage between organizational learning and performance. It can be noted that organizational outcomes are both time and space delayed hence there is often a weak linkage between organizational learning and performance governed by a phenomenon known as myopia of learning. Similarly ideas like environment, leadership and organizational outcomes are the competencies that can involve too many employees and their scope has too many outcomes. Similarly an organizations built in linkage provides sharing and learning of information and experience from one another. The content and capacity of learning are dominated by mission and vision. Again a linkage formed between learning, organization for competition and external environment creates competition.

Similarly another idea that is more in use is the Supply chain SC integration that builds the competitive strategy in identifying how three constructs influence firm performance through an SCM practice issue, the level of SC integration and competition capability and understand how this can lead to making the organization capacity much better. The presence of differentiation, diversification, flexibility and innovation identifies the success of a business in investments which helps in developing new products and pioneering new market opportunities after bringing stability to the economy and organization. Cost leadership priority can be connected with highly centralized and formalized administrative activities of the supply chain that can be linked to infrastructural and technological activities. SC integration impacts customer responsiveness and manufacturing performance via the key and most prominent linkage between sourcing and degree of manufacturing goal through a validation of this linkage. The power of integration in the development stage is that it can remove the barriers between functions and organizations and thus leading to strengthening of the competitiveness of the supply chain.

Linkages are built in various ways mainly that can create a relationship that can be scrutinized among 2,3 or more members. This relationship is a continuous flow that can get disrupted through outflow of knowledge from external environment. This knowledge may be created through outcomes and future or present processes of a business. But generally linkages are internal or protected by the hierarchy. Each element grows stronger through more positive feedback and creative and knowledge intensive input. Linkages can bring in competitive advantage in 2 ways: optimization and co-ordination. Linkages create a more costly product design, more stringent material specifications or greater in-process inspection. Procurement and reducing servicing costs are other parts that are related to product design that affect the manufacturing costs or cost of services and previous utilization. Linkages can be formed from any important core of a business. Channel linkages are similar to supplier linkages. Channels perform such activities as sales, advertising and display that may substitute for or complement the firms' activities. Exploiting linkages may also require the creation of switching costs as a by-product, tying one or both sides to the other. Trust and commitment are the key areas to exploit linkages. Emerging technologies and the already created research advancements in linkages have derived ideas that have made linkages easier to achieve. Computer aided designing and simulations have helped in creating other elements to create linkages from and use in deriving further business models as talked of in this chapter. Moreover it is linkages that have led to bringing interrelationships and technology advancements and cost advantages. Thus it can be said that linkages are a method of bringing together modern business concepts and creating results and benefits from the relationships of the subparts.

CHAPTER 10

Performance measurement as a superior best practice
(BY BAISHAM CHATTERJEE)

Performance measurement has an operational focus and a continual quest for service improvement. The main catalyst behind the benchmarking data is the findings from audit and findings from research. Performance measurement in the basic sense starts with defining the problem to find the evidence; and apply results for appraisal to evaluate the change. Other ways of measuring are by understanding audit to evidence; understanding from output to outcome to impact and increasing influence of risk management. Turnaround time is another means of creating speed of supply, supply time, response time and satisfaction of time-that measures the time between initiation and completion of a request. Request transmission and delivery methods are also subject to study in conjunction with turnaround time. Cost measures determine the amount of direct and indirect costs through research. Performance measures are based on location finders, number of requests, and growth in volume, fill rate, speed of delivery and document quality. Certainly, any study aiming to determine user perceptions of delivery speed must include the use of the measurement of different expectations. To look at getting the best output from performance measurement one should look at getting the highest benefit from investment and production. Companies at a certain point of time begin to lose market share to overseas competitors who were able to provide higher-quality products with lower costs and more variety. To regain a competitive edge, companies not only shifted their strategic priorities from low-cost production to quality, flexibility, short lead time and dependable delivery. They also implemented new technologies and philosophies of production management (i.e. computer integrated manufacturing (CIM), flexible manufacturing systems (FMS), just in time (JIT), optimized production technology (OPT) and total quality management (TQM)). The implementation of these changes revealed that traditional performance measures have many limitations and the development of new performance measurement systems is required for success. Performance measures have been primarily based on management accounting systems. This has resulted in most measures focusing on financial data (i.e. return on investment, return on sales, price variances, sales per employee, productivity and profit per unit production). The factors affecting the measurement of productivity include tactical, planning, strategic and internal management. Aggregate productivity measures account for all or most of the system inputs which requires significant amount of data that are time consuming and costly to obtain. Performance and productivity variables can be related as: Productivity is mostly concerned with direct labor which is no longer a significant portion of cost. Thus decreasing the cost of direct labor and/or increasing direct labor efficiency do not contribute significantly to the overall performance of the company. Moreover, focusing excessively on the efficiency of factory workers and departments detracts attention from improving the production system itself. There are many types of time based dimensions for measurement which are as, new product development includes: time from idea to market; rate of new-product introduction; and percentage first competitor to market. Decision making includes: decision cycle time; and time lost waiting for decisions. Processing and production includes: value added as percentage of total elapsed time; uptime yield; inventory turnover; and cycle time (per major phase of main sequence). Customer service includes: response time; quoted lead time; percentage deliveries of time and time from customer's recognition of need to delivery. Strategic analysis for performance measurement are as compilation of a set of data that facilitates a more comprehensive understanding of the key processes for the business (such as: factors that drive the overall demand for the company products, customer satisfaction criteria and performance against those criteria, the nature and strengths of competition, and the efficiency and effectiveness of the manufacturing systems). The analysis focuses on the most significant manufacturing features to develop forecasts of manufacturing process potential. A team from different disciplines in the company, such as engineering, operation, marketing and management, examine the potential of the manufacturing processes against historical data. The analysis has four main components: historical performance, theoretical limits, engineering limits and relevant benchmark information. Integrating these four components helps in developing three, five, ten year's potentials for process performance. For each process what if analysis can be conducted in order to improve the process and anticipate the effect of these improvements on the aggregate performance and understand the half-life concept and long time strategic concept. There is a dynamic performance measurement system that has the following characteristics: a clearly defined set of improvement areas and associated performance measures that are related to company strategy and objectives. This stresses the role of time as a strategic performance measure; allows dynamic updating of the improvement areas, performance measures and performance measures standards; links the areas of improvement and performance measurement to the factory shop floor. This is used as an improvement tool rather than just a monitoring and controlling tool; considers process improvements efforts as a basic integrated part of the system; utilizes any improvements in performance (i.e. going beyond just achieving improvement and actively planning for the utilization of benefits from an overall company perspective). It uses historical data of the company to set improvement objectives and helps to achieve such objectives; guards against sub-optimization; and provides practical tools that could be used to achieve all of the above.

Marketing process

The core of any performance measurement system will be those traditional, financially based performance measures which periodically summarize the organization's performance for the benefit of shareholders, lenders, creditors and statutory authorities. If managers were encouraged to make decisions based solely on improving such performance indicators, they would initiate short-term strategies aimed at improving bottom line results, perhaps to the long-term detriment of the organization. To overcome this deficiency in financially-oriented performance measurement systems several writers have recommended the implementation of a more broadly-based set of performance measures. Such performance measurement systems should allow managers to assess actions designed to generate sustainable, long-term improvements, but which may well have an adverse financial impact in the short term. It may be that a change in strategy simply involves re-focusing on a different set of performance measures on which data are available. In some instances, new and more relevant measures may need to be developed to facilitate achieving new strategic goals. Further, different strategic objectives must be accomplished within varying time periods so that the measures used to assess performance should be decided upon with this time perspective in mind. The context of the performance measurement system may be structured through its general environmental factors e.g., culture, legal systems, economy, management context, the history of the performance measurement system, and the system's network. We assume that resources, internal dynamics and context influence and affect the quality of a performance measurement system in different ways. Also, it is important to note that we can study a performance measurement system in various phases of its life cycle.

Performance measurement systems helps in monitoring and maintaining organizational control that makes the organization pursue strategies that lead to attainment of the possibly set goals and objectives. Improving quality and service is an idea behind performance measurement. In order to make an organization use these performance outcomes there should be a transition from measurement to management, which gives an effective strategic direction. There is involvement of leadership, communication between employees, stakeholders and customers in order to share assessment results. Compensation, rewards and recognition are related to results of performance. It is used in diagnosing systematic problems, evaluating effectiveness and links performance to consequences in order to strike a proper balance between risk and return. This system may have as management control system received great attention including short and long term ventures and helps the organization develop competencies, markets, and systems. By looking at measuring through this short and long term goals a routine organizational learning phenomenon can be brought out. Other questions that are to be answered during any performance analysis are the changing nature of work, increasing competition, specific improvement initiatives, changing external demands and power of information technology that helps in measuring performance. Business process re-engineering which is also a performance measurement tool calls for the horizontal flows of information and materials to be considered holistically when seeking performance improvements. This relies on a clear understanding of how the outputs of one micro process impact upon the next micro process, which in turn requires data to be fed back from the receiving process to the supplying process. Few organizations have measurement systems which allow this to happen, and rarely be the traditional measures of business performance appropriate. Measures such as labor utilization, for example, might provide some insight into how efficiently a process is running, but provide no indication of the impact the outputs of one process have on the next process in terms of quality and time. One of the first things organizations realize, when they begin to re-engineer their processes, is that once they have done so they will have to re-engineer their measurement systems.

A performance measurement system should be a dynamic system. Most organizations that have only a static performance measurement system should have a dynamic system but many organizations have only a static performance measurement system. The main barriers to an organizations ability can differentiate between improvement and control measures and develop causal relationship. Reviewing and reprioritizing internal objectives, deploying changes to internal objectives ensure that gains achieved through improvement programs are maintained. Brand business unit measures which consists of value for money and delivery reliability has operating process measures like sales, forecast accuracy, selling expenses, customer retention, ease of production, post-range new products and distribution costs, storage costs, product deterioration and delivery accuracy that matters the most in calculating process measures. There is always a logic for developing a simplistic scenario that can develop artificial intelligence techniques and look at the monitoring prospects that can decide the ability of the measurement plan. Neural network technology and dynamic performance measurement system look towards enabling the performance measurement platform through an ERP platform.

Logistics performance is based on effectiveness and efficiency where the major performance ideas depends on cost-efficiency, sales growth, profitability, job security and working conditions, product availability and on-time delivery where the key advantages are often easy and inexpensive to collect, and is likely to be comparable between organizations that can capture several important dimensions of performance often with impressive accuracy. For service measures such as order cycle time or lead time variability, the advantages and disadvantages are essentially the same as those of performance indicators. One limitation common to both is that there are many dimensions of performance which they cannot capture, particularly the extent to which customers are satisfied. Although there are several dimensions of logistics performance where hard measures cannot capture in a meaningful way, customer satisfaction is perhaps the most critical. A set of soft measures, collected using techniques such as the mail survey, telephone interview, or similar method are needed. Besides their usefulness in identifying problems, soft measures may also be called for where available hard measures are not comparable between one organization and another because of differences in accounting standards or similar problems. To think of the idea of performance measurement it is said that it ensures customer requirements, provides standards for establishing comparisons, provide visibility, highlight quality problems for monitoring the priority areas and provide feedback for driving the improvement effort.

The second component is of six types like innovation, learning process i.e. individuals knowledge and information system, resource utilization and benchmarking techniques. The measurement of performance is based on generating ideas with respect to market requirements, competitors' capabilities and best practice capabilities. It is supported by clustering of these ideas, look at structural and infrastructural facilities and then refining the vision on basis of group workshops and at last prioritizing these ideas. Through this, it is very important to look at the creativity, learning and imagination perspective. It is important to plan the benchmarking process by starting to collect data and information and enable the data by analyzing the performance gaps. Performance gaps and communication skills are highly related to control capability and other psychological skills that highly influence the job. These ideas are all related to effective leadership. In order to understand performance models, it is important to understand the business type model, competitive model, the life stages model that calculates the product life cycle. Similarly, the present stage model, organization adaptation model, financial crisis and uncertainty analysis model, technology change and management change paradigm model and new government regulation model are the key advantages that a firm should look at before taking any major step. A lot of research has been done on the cluster analysis based on the Porter s five forces and a business life cycle analysis like infant cluster, rational firm cluster and pioneer firm cluster develop a clear view on cluster analysis. Accountability is a key area to measure to look towards making all these ideas successful, they stand as a response and responsibility before superiors, subordinates and peers. Internal accountability has also to do with the responsibility for creating certain conditions that are deemed necessary for successful functioning of different units. External accountability has to do with meeting standards that pertain to legal, political, professional, economic aspects of institutional or individual behavior. Performance goals are needed to define the level of performance to be achieved by a program activity and express those performance goals in an objective, quantifiable and measurable form and describe the operational processes, skills and technology. It is important to establish performance indicators that are used in measuring and comparing actual program results. The different factors that can measure internal mobilization and performance by managing few concepts are as: appropriateness and measurement of the factors within the logistics strategic frameworks; relationship between productivity and performance to throughput management and customer requirements. Again, the other concepts are relevance and effectiveness of bench-marking of productivity and performance measures. Both internally and externally the applicability of the strategic concepts, as well as their process to specific industrial and commercial sectors, the issues and approaches needed for adapting strategic modeling for international operations and optimizing models for logistics are the key processes. The aims of this would be to ensure a balance between total productivity and consumer service levels.

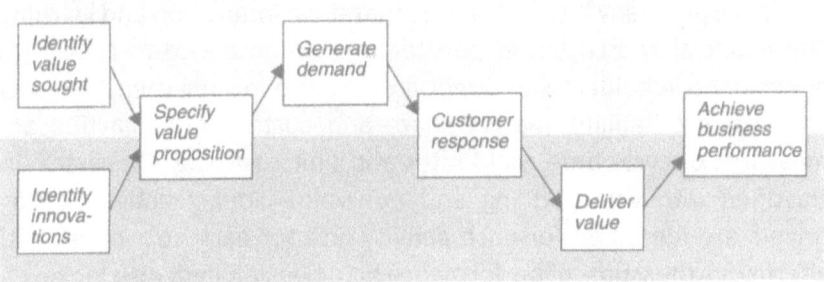

Domain of marketing

Domain of the rest of the company

Marketing to performance maesurement

The performance management process is the process by which the company manages its performance in line with its corporate and functional strategies and objectives. The objective of this process is to provide a proactive closed loop control system, where the corporate and functional strategies are deployed to all business processes, activities, tasks and personnel. The feedback is obtained through the performance measurement system to enable appropriate management decisions. In essence, the performance management process defines how an organization uses various systems to manage its performance. At the heart of the performance management process, there is an information system which enables the closed loop deployment and feedback system. This information system is the performance measurement system which should integrate all relevant information from the relevant systems. In this context integration means that the performance measurement system should enable the correct deployment of the strategic and tactical objectives of the business as well as providing a structured framework to allow the relevant information to feed back to the appropriate points to facilitate the decision and control processes. What is required most is integrity and deployment in understanding the environment consisting of market, customers, people, society, shareholders. It is used by setting direction, strategy, policy and objectives based on current performance that provides external intelligence with respect to strategic objectives of the business. The performance measures are used in understanding and developing the strategic plans of SBUs by understanding the market requirements.

There are many internal factors that drive innovation and learning. The financial and customer perspectives are intended to reflect the needs of stakeholders and target groups, and include measures such as sales, profitability, market share and customer satisfaction and measures of cycle time, yield rates and unit cost data. Processes are classified into value adding and non-value adding activities. Cost drivers are identified for each activity and for each cost driver with alternative measures of performance being designated. Also looking at the planning evaluation factors we see performance norms generated from, or in the course of, process planning. Measurement is based on latter evaluation process, also requiring contextual information. The latter includes the setting and external factors external to the process that have been identified as affecting performance. Performance measurement is based on the output of the particular objective. Outputs are the products or service that either flow from an organization to its environment in accordance with its mission, or flow from the process in support of the mission. Outcomes calculate the impact of output on stakeholder expectations. Everything can be calculated in the environmental, social, financial or on-financial dimension perspective. A strategic performance measurement system is the most talked of factor in the modern strategic dimension. It translates business strategies into deliverable results by combining financial, strategic and operating results by understanding the highest objectives of the firm and aligning management processes such as target setting, decision-making and performance evaluation. A part of the performance metrics in management process alignment where the main activity is designing and reengineering core management processes to incorporate new performance metrics as they evolve. The processes include: planning and capital allocation, performance assessment, management compensation and rewards and stakeholder relationship. Other processes are measurement and reporting infrastructure and establishing processes and supporting technology.

Performance measurement can be learnt upon and modified by researching best practices in performance and their causes, such as the use of supplier/customer contract relationships, self-managed teams, concurrent engineering, supply, chain management, etc. Stimulating regional strategies by which governments can begin developing an infrastructure that will attract networks of businesses; educating key players that productivity is the ratio between value added in the entire enterprise and the cost of resources consumed, rather than the narrow definition of output per man-hour, and that the key issue in this respect is engineering or re-engineering of business processes. Enhancing the blue collar human capital through appropriate training; shaping educational curricula for university students and professionals to prepare them for working in cross-cultural, process and team-oriented environments that stress collaboration within and between companies rather than the old adversarial lead to win-lose model of business. In certain situations where teamwork is required: challenging situations, such as a perceived crisis or outward-bound/team-building courses, which are completely different from the normal, daily, operational activities. In this type of situation, groups can progressively learn to become more effective at problem solving and work more as a team. Psychologically, it is often far easier to build up team spirit by making a clear distinction between the operational and change processes. The teams can then use performance information about the operational process to improve their problem-solving capabilities within the context of a distinct improvement process, even though the same group is involved in both processes. There are many ways of developing superior performance techniques which track development of productivity and to maintain low purchasing costs. In performance measurement it is important to track development of productivity, ensure availability of capacity and knowledge required to deliver solutions. It is important to track development of productivity, to ensure availability of resources. They should ensure access to economies of scale and access to components and services.

Existence of new forms of structures has been accelerated by the latest technological, political and demographic changes in the global marketplace. Business performance improvements are a result of close buyer-supplier relationship. The pre-contractual dimension refers to the factors that lead organizations to engage in factors that condition the selection of potential partners. These factors involve the necessity of an organization to acquire resources another organization which does not possess, the asymmetry between organizations. This implies the potential of an organization to exert its power over other organizations, the reciprocity between organizations which refer to the aim to have mutual advantage or the aim to share mutual risk, the drive for efficiency which refers to the objective of improving organizational performance. It also drives for stability which refers to the desire to minimize the effects of environmental uncertainty and the drive for legitimacy, to conform to the prevailing norms and values imposed by the institutional environments in which the organization is embedded. This is an example of pre-contractual dimension. To talk about performance measurement, it can be said that increased global competition, shortened product life cycles, accelerated technological advancements, and enhanced customer requirements have caused fundamental changes in the manner in which firms compete. Firms can no longer compete solely on the basis of price (cost), and must formulate competitive strategies defined by market-driven requirements. Therefore, it has become increasingly important for firms to develop strategic objectives which facilitate the development of a competitive advantage in specific markets or market segments.

Strategic objectives are initiatives designed to have a significant and favorable effect on the long-term health of the firm. The improvement of product, process and service quality has been adopted by many firms as a key strategic objective for achieving world-class performance levels. However, sustainable world-class performance will not occur if there is a misalignment between a firm's strategic objectives and actual market requirements. In addition, organizational coordination in relation to market driven initiatives is essential for ensuring the efficient use of company resources. In order for a firm to successfully compete on its strategic objectives, relationships must exist between the firm's strategies, organizational actions, and performance measure. Not only are specific action programs supporting strategic objectives required, but also integrated performance measurement systems which facilitate consistent organizational actions toward objective achievement. Performance measurement systems are composed of three elements: performance criteria, performance measures, and performance standards. Performance criteria are the relative elements used to evaluate performance. The actual values of performance criteria over some specified time period are the performance measures. Performance standards are the accepted levels of performance for each criterion. Moreover to realize the importance of criteria other than cost for the strategic success of the business the firm has incorporated quality, dependability, flexibility, and safety measures into their performance measurement system. As in the case of plant level control processes, division level planning processes may be accomplished in other ways than face-to-face meetings.

There are many important dimensions of the combination of value chain with performance, which consists of customer satisfaction indices, carry over index, have a number of changeovers, understand overlapping degree, defects after working cycle and understand project delay and prototype cost. It is very important to understand the machine productivity and adherence to supply schedules, vendor quality rates and stock level and rotation. The internal logistics and the outbound logistics consist of number of different models produced and zero-defect deliveries. All the different elements form the major advantage of SCM performance measures. Financial performance measurement is another important performance tool used alone is weak because market selling price is invariably the "driver" used to determine input/output conversion efficiency in the factory, while value-adding capability, cost of adding value and throughput time/ total factory cost are often obscured. Much criticism has also been directed at investment appraisal, yet it is still common practice to invest in isolated islands of efficiency in the production process without consideration of total material conversion constraints. Investment linked to total throughput time reduction within a value-adding framework is proposed.

A new paradigm of dispersed manufacturing is emerging. Whether or not a company is able to compete in the local and global marketplace depends heavily on how effectively the company can build up the partnership synergy with their partners which are also service providers. In particular, it is essential to develop the necessary techniques to measure the performance of business partners in terms of delivery lead time, quality and price. Not having a reliable means to assess how good or competent a business partner or service provider is prior to collaborating on a business project is obviously an unwise move. In the existing practice, the assessment method is relatively subjective and normally based on personal views/experience together with some recent business records which cannot reflect the actual picture of the overall company performance. Without looking not only at the manufacturing paradigms and also concentrating on the design processes at first the firms mission statement can be clearly defined moreover a firms strategy objective can be defined by developing on global performance and functional areas role. It is also important to communicate strategic objectives by identifying the supply feedback and the competitive position. To derive a true value of the performance measures, it is important to group products, agreeing performance measures for the business objectives and embedding and signing off the top level performance measures. To have a different outlook, the decisions made at different levels of the organization vary in nature, but they should all strive towards the same overall strategy. Increased focus on quality, dependability and flexibility, and the fact that strategic priorities might vary between products, and between stages of a product's often short life-cycle, sometimes make it hard to link measures to strategies. Performance measures may even hurt a company's corporate strategy due to mismatch between goals on different level. Considered that qualitative and non-financial manufacturing performance measures can help organizations to link operations to strategic goals on all hierarchical levels, since they are easier to derive from the qualifying and order-winning criteria and easier to put into effect, but it is still necessary to link corporate, business and manufacturing strategies. Looking at the performance rate measure, and its relation with production time, it can be described as. Planned production time (or loading time) is separated from theoretical production time and measures unplanned downtime in the equipment, i.e. by this definition unavailability would not include time for preventive maintenance. This definition gives rise to planning of preventive activities, such as preventive maintenance, but it might lead to too much maintenance of the equipment and too long set-up times. If planned downtime is included in the production time, the availability would be significantly lower, but the true availability would be shown. That would create motives for decreasing the planned downtime, e.g. through more efficient tools for set-up and more efficient planned maintenance. The performance rate measures the ratio of actual operating speed of the equipment i.e. the ideal speed minus speed losses, minor stoppages and idling and the ideal speed based on the equipment capacity as initially designed measures a fixed amount of output. It indicates the actual deviation in time from ideal cycle time, on the other hand, focuses on a fixed time and calculates the deviation in production from planned. Both definitions measure the actual amount of production, but in somewhat different ways.

By looking at team effectiveness in performance measurement, the improvement in team responsibility was most notable in devising internal measures that shifted the focus from labor productivity to the cost effectiveness of operational activities. This provided an incentive to reduce costs and a system for assigning responsibility for controllable costs. A renewed focus on product quality and customer focus emerged from resource consciousness in the team development process. Also significant was the role played by the internal team cost coordinator in a business who leads the implementation of the key performance measures established in the activity-based management project and was establishing links between materials management and cost control strategies. A heightened focus on team performance creates a greater chance that teams may make dysfunctional decisions or perform in a manner which improves team performance ratings but is counter to overall corporate performance. This may be overcome by increased training focusing on organizational values and the use of broad based performance measures that demonstrate the interdependence of teams. Management accounting should be able to provide the essential links between teams and the organizational co-ordination needed to prevent any sub-optimizing behavior from teams.

Product planning translates the product concept into specifics for detailed product design including styling, layout, major specifications, cost and investment targets and technical choices. Interior and layout evaluations and early stage prototypes which interprets the product concept in the physical form. Similarly the contrast between volume producers and high-end specialists is also manifested in the way they perceive markets and differentiate products. Cost, profit margins and price create the major performance gaining focus. Thus performance measurement can bring about huge changes through the ideas that is provided here and focus on product development and other internal development strategies.

CHAPTER 11

Competitive benchmarking: a correlation to internal competitiveness and ability

(BY BAISHAM CHATTERJEE)

To see to it that the level of economic experience, the sustainability of research and development and investment in human capital staffs afloat a business should look at competitive benchmarking. All businesses at this point should attain peak performance in their product delivery process. A best in class standing of the world market is benchmarking. Logistics is an area that needs development and exceptional interrelationship, linkage and value chain modification skills with time. Competitive benchmarking has developed a process or product that is more effective and superior to anyone else. Performance measurement and continuous monitoring is a benchmark for the category of processes. Benchmarking roles in logistics brings competitiveness and hence develop the process model. Benchmarking means comparing practices and results with the best organizations in the world, and then adapting the key features of those practices to one's own organization. Another important idea is accelerating organizational learning, customer-driven quality and continuous improvement. It is also important to help organizations identify breakthroughs, by comparing their processes to those of the organizations recognized as being the best. It is also important to help organizations learn from each other whether it is in business, health care, government, or education. The goal of benchmarking is ultimately to learn and incorporate process and product innovations that have proven successful in other organizations. The emphasis in benchmarking is not on the outcome, but on the process employed to achieve an objective. Thus, benchmarking process facilitates the transfer of technology. Leading experts have expressed clearly the opinion that in today's competitive business, organizations that are characterized as dynamic in their learning and adaptation process will ultimately survive and thrive in a highly competitive market. Benchmarking provides a systematic process to learn and adapt an organization to the best business practices. The operations of a firm have been described as a product delivery process, a value-added chain, a system or network, and a conversion process. Segmentation and functional specialization have been cited as key contributors to the loss of competitive performance on an organizational level. Optimizing individual functional performance has often led to sub-optimal corporate performance. Integration of functional activities in a manner that supports the corporate strategy is viewed as being necessary for sustained competitive advantage. Benchmarking effectively provides the necessary link between functional performance and the corporate strategic position. It serves to identify sources of competitive advantage that can be exploited more vigorously to provide company-wide areas of distinctive competence; and it helps uncover weaknesses that need to be eliminated to become more competitive in the marketplace. Researching on benchmarking ideas needs various success factors. This develops on ideas like process comparison and monitoring process performance and performance gaps. This determines the idea that leads to performance gap that ultimately leads to innovation gap. Similarly benchmarking has relation with the best practices that modify company environment and lead to business process improvements. They defined the scope of their benchmarking effort to include the entire product-delivery process. The goal of this effort was to improve the order fulfillment process by becoming more responsive. The company's competitive position in the marketplace has been enhanced through the benchmarking efforts. This supplier has been able to improve its responsiveness to the customer by reducing the cycle time. The benchmarking process was organized as an individual project, with a steering committee and a project group, both comprising members from all the major actors in the supply process, such as planning, operations, services, technology and quality. When finding gaps in performance and practices in a sub-process, another round of detailed critical examination of that process may be necessary to understand the differences better. Such in-depth analyses may lead to a more detailed description of the process, as well as more detailed performance measures. This is an important part of actually understanding the differences. When finding gaps, explanations are needed. Therefore, various diagnostic measures are called for in a benchmarking study. Diagnostic measures help in the comparison process.

Process planning is another important identification to understand the criticality in cost savings potential and customer satisfaction and the other direction that consists of process mapping and performance metrics consists of document process flow and performance data. It is important to identify gaps and analyze the feasibility and then integrate top management with the team. To look at the basic characteristics of competitive groups, it can be identified that the notion of the primary competitive groups makes the concept of strategic groups more complete because rather than categorizing these firms just from the supply-side of their operating environment as the strategic group theory does, the notion of primary competitive groups also includes marketplace and technological characteristics since these features are interrelated in the minds of managers. Consequently, the primary competitive groups represent the psychological reality for their members. There is a three dimensional framework SRP or strategic reference framework involved in the initial theory of competitive benchmarking. It argues that previous investigations have only used industry averages as a point of reference, whereas the SRP matrix considers simultaneously multiple points. Specifically it looks at three dimensions; internally to the firm's aspects (strategic inputs such as targets and capabilities and strategic outputs such as profitability and performance); externally to the firm's conditions (customers, shareholders and competitors); and time (past, present and future) orientations. The SRP model is regarded as complementary to, or even the missing link, in the structure-conduct-performance (SCP) paradigm. Specifically, it is postulated that decision makers' reference points, or benchmarks, along the three dimensions of the SRP (internal, external and time) influence competitive strategy development. And, if according to managers' perceptions the company is in a worse-off position relative to competition, they formulate risk-assertive strategies whereas if the company is in a better-off position they formulate risk-aversive strategies. To look towards making the benchmarking ideas better a business should look at making senior management support better, making ideas open to change, help in developing benchmarking partners and develop efforts to make this benchmarking successful. Moreover to formulate the first seeds of competitive benchmarking the business should not discuss about competitively sensitive costs or sensitive data.

Competitive-cost benchmarking is an action-oriented tool that enables companies to quantify how their performance and costs compare against competitors, understand why their performance and costs are different, and apply that insight to strengthen competitive responses and implement proactive plans. Benchmarking, by definition, goes beyond competitive-cost analysis, which is often a staff exercise without a structured follow-up implementation program; its goal reaches beyond simple competitor emulation. The profitability of manufacturing operations with a given technology and cost structure depends on volume, price, and product mix. Pricing of commodity products is competitively set on a cost-plus basis (not on the basis of the value to the user). The profit margin depends on industry capacity utilization and on the economics of the price leader, usually a quality producer with the lowest costs. Organizations that develop a detailed qualitative and quantitative understanding of competitor and competitive costs are more able to define and implement strategies that cope with industry realities. Significant bottom-line improvements can be gained through the competitive insight thus acquired—whether by copying competitors, adapting superior practices, cutting costs, modifying pricing policies to exploit competitor weaknesses, or committing to new capital investments or acquisitions. Normally, commodities are characterized by continuing cycles: feast or famine. During product shortages, even high-cost producers may enjoy acceptable margins. But during a downturn, producers with lower costs are in a stronger position. They are able to increase their business at the expense of higher-cost producers, possibly forcing some to go out of business and setting the stage for the next period of balanced supply and demand and acceptable profits.

Edward Deming's ideas from which competitive benchmarking came into practice the best organizations in the world need to adapt to the key features of those practices to one 's own organization, as well as it is important to accelerate customer driven quality and continuous improvement by creating breakthroughs from the processes that are recognized the best. This type of benchmarking helps competitors and collaborators learn from one another. Segmentation and functional specialization are key contributors to growth or loss of competitive advantage. Hence this has to be managed very fruitfully. Optimizing individual functional performance has often led to sub-optimal corporate performance. Benchmarking effectively provides the necessary link between functional performance and the corporate strategic position. It serves to identify sources of competitive advantage that can be more vigorously exploited to provide company-wide areas of distinctive competence; and it helps uncover weaknesses that need to be eliminated to become more competitive in the marketplace. The benefit of this approach is that it aligns the operational activities on the functional level with the overall needs of the corporation. It is thus capable of providing the strategic focus at the functional level. Planning and organization are other divisions that lead to benchmarking like criticality, cost savings potential, customer satisfaction enhancement and management support that set the perfect background for benchmarking. The willingness to allocate time and resources is a key example for setting relationship between data analysis and process reengineering after the process of data gathering is done. All the gap analysis perceptions are determined during this phase, which helps in understanding the root cause behind these gaps and gather adaptability to create a more integrative and redefined process.

Competitive benchmarking can be followed by the analysis of pioneering firms. Although pioneering firms may obtain high rewards due to the absence of competition, their strategy of being first to market can involve high development costs as well as effort spent educating the market on the pioneering new product's benefits. Even though some followers may be aware of the same new product market opportunities as pioneering firms before the pioneers actually enter with their new products, delaying their new product introductions until after the pioneering entries also reduces cannibalization of the follower's current products. If follower firms delay too long in their response to the new product introductions of pioneering firms. However, they risk the possibility of being overtaken by faster follower firms who will benefit from introducing their product into a viable, less-crowded, and less-competitive marketplace. Awareness helps in understanding environmental monitoring that helps in developing new product through a very efficient market study. A firm should have interest in strategic decision makers who have key ideas in boundary spanning that helps in making product market developments much better. It is important to understand internal product development on a competitive new product and develop coordinated efforts in developing, testing, and launching the new product to reach the pioneering competition. Launching a new product needs organizational decision making based on test market performance and information on the product market. It is important to scan the environment for new product market opportunities and understand the pioneering developments of a related product market, before the introduction of a new product. It is important to monitor strategic decision makers and understand the ideas of new/current product developers. Developments in the market can be carried forward by understanding the scanned environment that has understood the proactive nature of the early entrant that can create an opportunity. Later entrant firms can gauge the competitive response performance against firms that are proactive in search for identifying, responding to, and exploiting new market opportunities. Later entrant firms can improve performance most in the early stages of competitive response, yet many can also improve their performance in later stages although to a lesser degree. Competitive benchmarking can be identified by understanding performance measures for each function of a business operation. The easiest way is by understanding a company's internal performance levels as well as making a competitor analysis and evaluates the performance levels and implement programs to close the gap between internal operations and the companies. To take more important decisions, we have to understand the marketing opportunities that involve providing a brand or corporate identity, branding of service sector and relationship of the corporate identity with service branding.

Moreover professional standing of the organization and its relationship with the industry; look at the fact that the change of corporate attitudes and policy statements can be understood with time. It is important to understand the relative size and position and company infrastructure. The quality performance and maintenance of standards come through training, expertise, experience and continuity. The customer base can be understood through major contracts, new contracts, growth and innovative services.

Competitive success as a nation requires balancing commercial innovativeness and social welfare, which results in a sound basis for socio-economic development. All potential resources including entrepreneurial activity and innovations can be utilized as promoters of competitiveness and welfare. In the context of national competitiveness, benchmarking seeks to increase national performance through improved policy design and practices. This need has been more evident with the openness of world markets, growth in international trade and the globalization of industry. For a business to gain future competitiveness manufacturing strategies under the dynamic and complex situations relies on forward thinking strategies. Businesses utilize a multi-focus manufacturing strategy with a lot of ideas in the business plan that would enable the recruited talent drive all the organizational and technological processes ahead. Competitive circumstances determines the nature and ranking of the business, the other one is the priority of innovation and new or prototype product development of the business and the third idea is what are the things to be done to bring the plan and objective to a success by measuring the circumstances. Benchmarking analyzes the nature of a business and the ranking the business can be brought to base on whether the firm is a prospector, analyzer or defender. The competitive priorities of manufacturing is based on know how, flexibility, delivery, quality, customer focus and costs, where the sustainability of innovation depends on creativity and adapting to change and changing the nature of product modeling and value chain according to the nature of competition and ultimately look at product performance and costing methods. Thus this is a very normal phenomenon which can be applied in any country, in any growth nation depending on the previous performance and capacity of the product to win in the market. The different organizational qualities that drive growth are process reengineering, restructuring, organizational focus and developing efficiency and efforts in teams. Channel power towards retail level, looking at customer service, strategic planning and differentiated strategies. Activity based costing and alliances help the business after the goal is set based on performance comparison. It is important to identify customer relevant terms based on the mission and functional roles that might add value to the integration and performance effort required to bridge traditional boundaries.

However, confusion as to what benchmarking actually is has resulted in many companies using results or cost-driven approach, the objectives of such benchmarking being to achieve purely a cost reduction. This involves them comparing some aspect of their performance with those of superior performing competitors, usually using some intermediaries such as consultants. This approach is characteristic of the short-termism in companies that cite strong financial management as their major business strategy. Competitive cost-driven benchmarking has been the most common type of benchmarking carried out in the West, and has been a major catalyst in helping companies reduces costs. However, it is somewhat surprising that cost has been the main emphasis of benchmarking efforts, considering in most markets product differentiation, rather than low-cost production, will determine success or failure. Process-driven benchmarking when applied to processes that effect product differentiation, can enable companies to improve dramatically their competitive position by enhancing areas such as product features, quality, new product development, reliability, customer service brand image, etc. It is important to redefine competitive gaps in cost, quality and timeliness by the outcome and measures of breakthroughs and measuring competitive parity and increasing advantage. Internal benchmarking is very common with companies comparing productivity from one time period to the next. Whether benchmarking does reflect improvements due to facility layout change is less certain. Companies are trying to become more competitive and attention is paid to such analysis. A common improvement that companies are trying to make is that of improving materials handling and stock control. In the last few years there seems to have been a major trend towards the use of more automated warehousing and towards the general reduction in total stock. In terms of facility layout, it would appear that facility improvements have been most prevalent in this area, i.e. warehouse facility layout. To have a clear view of competitive benchmarking, there is necessary to trace costs of products, processes and activities. It is important that non-value activities are not isolated and the cost of quality has to be identified based on customer service, flexibility and throughput. It is important to have logical ideas like profound process knowledge, innovative thinking and develop self created solutions for problems that are innovative. This is the perfect way to start a benchmarking process.

Benchmarking has an initiative through pinpointing the firm's strategic location, pinpoint the competitors' strategic location, shifting to a different dimension with time and plan and implement strategies that compete directly and win, hold a position, negotiate a settlement and regroup and change a basis of engagement. The strategic landscape for firms comprises indicators such as their relative competitive position, the industry conditions under which they and their competitors operate, and the core strategies that are being followed by themselves and their competitors. This is, of course, just a foundation for strategic success and not by itself a sufficient condition. Thus success quite often is unlikely without original or entrepreneurial strategies which chart the firm's direction and steer it clear of likely strategic impediments to success. This often comprises information on issues such as: details of major competitors' goals, strategies, scale etc.; environmental influences; internal competencies. For example generic factors such as: relative market share, relative product quality, proprietary patented processes and products, brand power, market growth rate, industry concentration, capital intensity, cost and investment structure, buyer and supplier power, new entrant threat, etc. all remain timeless strategic determinants of performance.

Another idea of competitive benchmarking is known as environmental benchmarking. The integration of environmental benchmarking in this business excellence model has created one of only few examples so far where a multinational has succeeded in structural integration of environmental performance criteria into mainstream business criteria. As such, environmental benchmarking has been highly successful in generating environmental improvements for numerous products, but has also provided eye-openers for cost reductions and opportunities for innovation outside the environmental context. In process level benchmarking, it is easy to make benchmarking a continuous process, due to the relative smallness of a single benchmark study. As this will result in benchmarking of consecutive models of a certain product it allows for trend research. It provides a level of detail otherwise unobtainable. Solutions that can be easily and directly implemented will be identified. Such low hanging fruits would probably be missed in the big picture of a process/strategy benchmark. It is easier to distribute costs. Costs can be easily redirected to business units as costs per project are relatively limited. It is easier to select benchmarking partners. As the benchmarking is concerned with a product of a competitor instead of their organization they do not have to actively participate, which might be difficult from a competitive or even anti-trust point of view. An organization should look at ways to create new knowledge and seek for rapid organizational learning-unlearning and innovation where all the processes are dependant on technology, procedures which all required source and a generalized plan and thought process to fill up the knowledge gap. The benchmarking process, as intended, is a learning process that involves observation of external practices and performances, comparison with internal ones, identification of knowledge gaps and finally the decision: bridging the gaps acquiring new resources or leveraging on internal ones and investing in upgrading. The result of this learning process is something new; deriving both from the integration of external inputs with internal previous knowledge as well as from bridging knowledge gaps. Benchmarking requires various analysis phenomena to make the performance better. A pre-analysis of ways to benchmark ideas requires identifying performance drivers, planning and data gathering that at last has to be modified in respect to the constraints. After the pre-analysis phase, comes the analysis phase where common metrics, performance gaps and any other success factors and internal factors are related by solving case studies. But case studies are not always required if the content is through new idea generation. Project designs and mapping and external resource utilization analysis are automatically linked to identifying competency gaps that provides the main source of decision making to reduce the knowledge gap. Reducing the knowledge gap requires using creativity and talent. Strategic benchmarking, which is a part of competitive benchmarking have practices that follow this direction, first compares the company's current products and services against customer needs and expectations. To begin, a focus group of company employees identifies products and service factors they believe are rated highly by their customers. Once the criteria that describe the current product and service practices have been determined, a survey measurement instrument is designed. Each identified criterion is incorporated into a research question, then rated by customers into two dimensions. First, customers rate how important the criterion is to them to be satisfied with the product or service as provided by the company. Next, customers rate how well the company is performing against that practice criteria.

There are a lot of benchmarking ideas attached to benchmarking in the organization and establishing knowledge areas in the industry. This is followed by the strategy planning, strategy implantation and strategy evaluation that have gaps in the targeted company that helps in establishing a strategic matrix. It is important to take competitive strategies into action and then examine the competitive strategy. The proposed effective strategic management tool, the strategic matrix, is established in order to assist firms to re-examine their competitive position. Four strategic cells are formed: core competences with low gap; core competences with high gap, non-core competences with low gap and non-core competences with high gap, leading to the formation of four strategies, "anchoring", "narrowing", "catching-up" and "following-up". Even though Matrix approach is not a new method in analyzing company operational strategies, however, the strategic matrix we proposed incorporating the concepts of environmental analysis, benchmarking and the customer's voice has never been used in previous studies. The most noticeable factor is that in a volatile world in which the identity of customers, their preferences, and the technologies for serving them are all changing, a market focused strategy might not provide the stability and constancy of direction required as a foundation for long term strategy. Resources and capabilities have a much more stable basis upon which to define a sense of identity. Tangible resources are a source of gathering knowledge, motivation and competences. Core knowledge improvement is related to organizational learning capabilities that help in gathering competences in the same business segment. Monitoring and evaluating are of various types, starting from performance measurement value creation, intangible assets value creation, and activity based view and knowledge, redesigning knowledge. The source of profit generation and wealth creation has to be understood and it can be better understood through the stock flow theory that brings in growth, renewal, efficiency and stability. Benchmarking is more than a comparative analysis of financial and technical measures from one company to another. To be effective, benchmarking activities need to be integrated into organizational strategy and the process needs to employ a broad range of balanced performance measures consistent with organizational strategy. In this new integrated system and in this idea of benchmarking, organizations need to accept a long-term perspective and to utilize balanced, broad reaching financial and nonfinancial performance measures to carefully improve the competitiveness of the entire organization. This approach requires that benchmarking organizations develop a complete understanding of their own business strategy and deployment of the strategy into functional strategies. This process will ensure that there is a consensus within the organization about long term and short-term performance measures that are consistent with organizational mission and goals. Action plans is a significant idea behind benchmarking. For traditional companies, manufacturing action plans will primarily focus on quality improvement activities followed by linking manufacturing strategy into corporate strategy, manufacturing reorganization, and employee related activities. While the emphasis on quality improvement activities is a sound action plan, the order of the last two activities should have been reversed. For JIT companies, manufacturing primary action plans will focus on manufacturing reorganization, linking manufacturing strategy into corporate strategy, developing collaborative supply chain, and integrating information systems into supply chain. Focus on a number of employee related issues such as employee training, empowerment, teamwork, and fair compensation followed by a set of quality improvement activities are the next series of manufacturing action plans. The ranking for JIT organizations are more consistent with organizational goals and objectives and with the principles of JIT systems.

To look for future competences and to take benchmarking ahead best practices can be incorporated into a company's operations. It also helps in providing realistic target and resistance and understands technical ideas and finds the knowledge base. It is important to focus on competition to evaluate the gaps that may exist. The other processes of benchmarking are, whenever reengineering takes place, there will be individuals on the team who will be skeptical about certain aspects of the new process, whether it be technological feasibility, cycle time reduction targets, or organizational structure. It's important and advantageous to include these skeptics in the benchmarking activity. First, it allows them to see first hand how another company operates in what could be their "future" environment. Second, their skepticism can be used to balance the views of the group when discussing the applicability of a benchmarking partner's approach. The most important challenge is to select benchmarking partners with truly innovative approaches. This will encourage the members of the reengineering team to let their imaginations run free. Additionally, developing an understanding of how the benchmarked organization managed change can be invaluable for your own implementation strategies and tactics. Effective benchmarking is characterized by an organized marketing intelligence system that is critical in selecting benchmarks and in evaluating the trends of important, designated benchmarks. Thus benchmarks can be seen as an integral part of forward looking management, competitive intelligence and the analysis of changing markets. They can provide stimulation and motivation to marketing executives whose creativity is needed to execute the benchmarking activity. Benchmarking is a mechanism for inducing inter-industry change suggests that benchmarking breaks the ingrained reluctance of operations to change. People may be more receptive to new ideas and their adoption when they do not originate in their own industry. Benchmarking may also identify a technological breakthrough (that might not have been discovered in one's own industry for some time to come), such as bar-coding, originally adopted and proved to be effective in the grocery industry. In these instances it is more important to uncover the industry's best practices than to concentrate on obtaining comparative cost data. Benchmarking means a continuous improvement through business excellence. It is important to understand learning organizations and adapt to information sharing. Competitive benchmarking is based in various ways relative to time and necessity in adoption starting from calculating the production lag, enabler benchmarking to reverse benchmarking and generic benchmarking. For example competitive benchmarking in case of electronics firms is based on major competition from low cost far east competitors; rapidly changing technology; move towards higher value added products; continual demand for new products with short development timescales. Similarly in the case of electrical utility competitive benchmarking is used in regards to privatization and deregulation; rapidly changing market; new entrants with less overhead; program of diversification and investment.

There are many processes for processing the value chain in benchmarking which consists of incoming orders to material management to invoicing in case of administrative processes and demand forecasting to packing and distribution planning in the planning process. The principles of benchmarking are equally valid for large, medium or small companies, whether they compete in local, national or international markets. However, the aims and achievable objectives of the benchmarking process may be significantly different in each situation. For example, one company may benchmark in order to improve profitability by identifying the most cost-effective way of maintaining their current standards, while another company may benchmark to identify industry best practice. For companies aiming to achieve recognition for world-class performance, it may be necessary to move away from a purely reactive approach to benchmarking, which identifies what other groups or companies have already achieved, and move on to essentially a proactive approach by identifying and defining the highest standards achievable. Thus benchmarking requires a well defined and creative correlation of all benchmarking elements and then research on the complete idea and then make an experimental application to reach the ultimate decision.

CHAPTER 12

The necessities and importance of using analytical hierarchy

(BY BAISHAM CHATTERJEE)

It is important to understand the analytic hierarchy process and the application of the analytic hierarchy process (AHP) to measuring and comparing the overall performance of different manufacturing departments on the basis of multi-attribute financial and non-financial performance criteria. The AHP is a multi-attribute decision tool that allows financial and nonfinancial quantitative and qualitative measures to be considered and trade-offs among them to be addressed. The AHP is aimed at integrating different measures into a single overall score for ranking decision alternatives. Analytic hierarchy is described in its usage in various ways to develop a hierarchical structure of the decision problem in terms of overall objective, criteria, sub-criteria and decision alternatives and determine, on pair-wise basis, the relative priorities of criteria and sub-criteria that express their importance in relation to the element at the higher level. It is also important to attribute, on pair-wise basis, the suitability ratings of the decision alternatives with respect to sub-criteria. Moreover there are many operating measures in manufacturing. conformance rate (percentage of products that meet the customer requirements), inspection costs (expressed as a percentage of the overall production costs) and a qualitative measure of the rationalization degree of the operating procedures are the quality related measures; flexibility is measured by product flexibility and technology flexibility, which are qualitative measures of the possibility to introduce at low cost respectively new products and new operations, and by volume flexibility, a qualitative index of the capability to make rapid volume changes; environmental compatibility indexes are: solid waste (measured by the ratio of tons of waste to production volume), energy consumption (measured in fuel oil equivalent tons per product) and a qualitative measure of the green image of the factory in relation to local people and institutions.

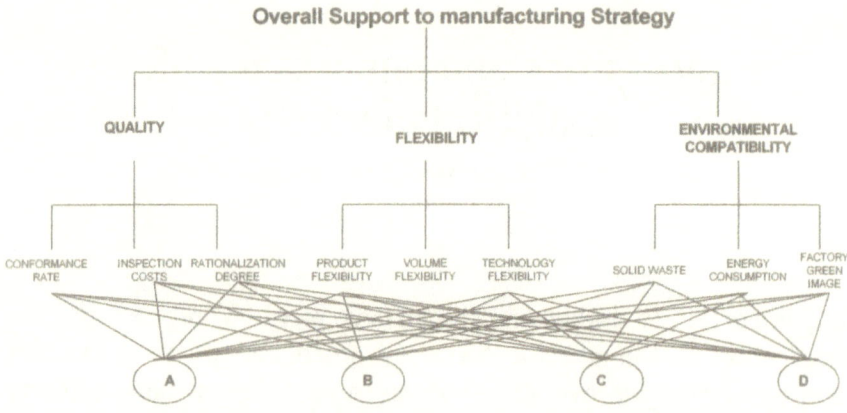

Fig.12.1 : Performance Hierarchy (A. Rangone, IJOPM,1996)

In Fig 12.1 conformance rate, flexibility and environmental compatibility are compared based on the qualitative aspects based on the hierarchy level of four factories: A, B, C, D. It is based on financial measures, non-financial physical measures and non-financial qualitative measures. An AHP helps managers in encouraging discussion among managers about the manufacturing competitive priorities and performance measures, facilitating managers to reach agreement on those that are critical for the company. Help managers to communicate the manufacturing competitive priorities and the relative importance of performance measures to all levels of the organizational structure of manufacturing, by translating managers' subjective judgments into quantitative terms. This ensures congruence between individual actions and manufacturing strategy. AHP helps in understanding performance measures by standardizing to organize the control procedure, avoiding their dependence on subjective and implicit judgments of managers and management accountants; understand better the overall support of each department to manufacturing strategy—and, thus, ultimately to the achievement of the overall business strategy—by addressing trade-offs among different performance measures (or competitive priorities). AHP differs from other multi-criteria methods in its ability to identify and take into consideration the decision maker's personal inconsistencies. Decision makers are rarely consistent in their judgments with respect to qualitative issues. The AHP technique incorporates such inconsistencies into the model and provides the decision maker with a measure of these inconsistencies. A consistency ratio is derived from the ratio of the consistency of the results being tested to the consistency of the same problem evaluated with random numbers.

Moreover, AHP can access simulation software packages. Simulation focuses on the formulation and solution of problems by trial-and-error methods. The modeling process in simulation is iterative and often reveals important information and new insights into the problem area as a result. During this iterative procedure the relationships between the system under study and the model are defined and continually redefined. Simulation and modeling are therefore inextricably linked with the steps in the simulation process, these being to: formulate and understand the problem; specify the model; build the model; simulate the model; and use the model as an aid for decision making. In order to accomplish a successful simulation project encompassing the above five steps it is imperative that the simulator has experience of the modeling environment and can develop simulation expertise in relation to the stimulant, the model and the simulation software. A prerequisite for effective use of simulation facilities would therefore be that the simulator be an on site member of the user organization's resource. It is also essential that an appropriate simulation package is selected as it can have a big impact on the ultimate validity of the model and on the timeliness with which the project is completed. Ideally, any proposed simulation software package should exhibit the ability to perform satisfactorily in all three decision-making categories like strategic control, management control and operational control. Indeed, most packages are tried and tested in relation to strategic decision making and recently there has been a shift of emphasis in simulation from system design to system control. The use of simulation as a real-time scheduling tool is becoming the subject of a number of feasibility studies and managers are beginning to look to simulation for an aid to day to day operational problems. Internet advertising has a proposed AHP model which is as: initially as uses in group interview of experts to identify suitable evaluative criteria including qualitative and quantitative and quantitative criteria for selecting internet advertising networks, after which AHP is applied to determine the relative weights of the criteria, rank the alternatives, and thus select the ideal internet advertising network. The AHP model can also combine both qualitative and quantitative criteria, making it an appropriate approach for solving the current problem.

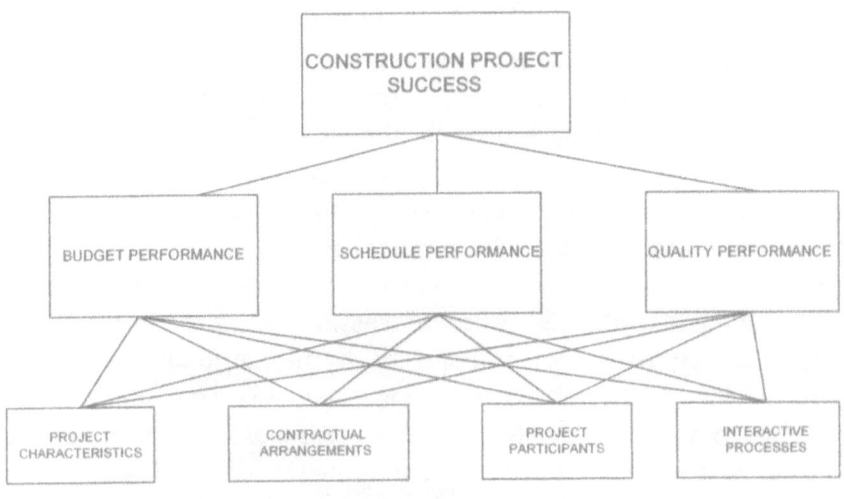

SOURCE: Cheng & Li (2002) ; Chua et. al. (1999)

In Fig 12.2 the first level is the decision problem which created 3 performance objectives, which has been sub-divided into 4 elements. AHP is known as an essential tool for practicing managers and academic researchers to conduct research for making business decisions and examine management theories respectively. However, the underlying concepts of AHP are not inherently built but are mastered by the users through courses of self-learning. It is said that

AHP is a multiple criteria decision making (MCDM) that has been used as a quantitative tool of quality function deployment (QFD). It consists of tailor-made questionnaires to collect the perceptions from experts or decision makers that scrutinize the groups of elements through a framework that consists of groups of elements of rating. AHP is referred to as the most powerful and widely used technique for decision making. It allows decision makers to measure the consistency and stability of their decisions. AHP has been shown to be useful in prioritizing alternative variables. AHP has been used in a wide variety of complex decision-making problems, such as the strategic planning of organizational resources, the evaluation of strategic alternatives, and the justification of new manufacturing technology. AHP has been applied in a variety of formats such as: a design tool for large-scale systems or composite ratio scales, an instrument for pair-wise comparison in the application of artificial neural networks, or a primary structure of decision-support systems. AHP allows a number of individuals and groups to participate equally in the decision-making process. The analytical process can provide a critical link for developing trust and true group participation. AHP allows diverse viewpoints to be considered and integrated ensuring that all participants have input to, and ownership of, the final evaluation. Also, it is important to note that the decision making process is rational, consistent, and defensible, aiding objectivity and understanding. It is assumed that all criteria are independent. This precludes interactions among the criteria. AHP assumes that a decision maker can compare two alternatives on a specific criterion without considering the other criteria. It clearly structures the process of establishing priorities. Although AHP presumes that the phases are executed sequentially, the decision maker can return to a previous phase in order to make changes. Because of the flexibility of the method it is not necessary to repeat all judgments when a change is made. Changes in the model, such as the addition of an alternative or a new criterion have only a limited impact on other parts of the model.

The AHP process has been applied in a variety of decision making and analytic arenas. It has a wide variety of operations management problems and decision frames as the AHP model is often able to satisfy the complexity of operations management problems without oversimplification. Several applications of AHP including supplier selection, facility location, forecasting adjustments, technology selection, product design, plant layout, maintenance scheduling, and logistics carrier can be identified. It is important to utilize an AHP model to identify a firm's core competencies using both quantitative and qualitative input. An AHP application for supplier selection as well as well can be outlined, citing AHP as a good tool to enable consideration of, and resolution to, the inherent conflicts between decision criteria, i.e. cost and quality. (AHP) is proposed as a suitable technique for analyzing information systems outsourcing decision. AHP can be used to rank all possible alternatives of outsourcing using several criteria. It converts subjective assessments of relative importance into a linear set of weights that can be used in ranking the alternatives. With this technique, several options are considered in the decision analysis that make it possible for a company to adequately evaluate and determine whether particular IT should be outsourced. This approach may be helpful to the practitioners who are faced with outsourcing task.

The process of AHP comprises the following steps:

1 Structure a problem with a model that shows the problem's key elements and their relationships.
2 Elicit judgments that reflect knowledge, feelings, or emotions.
3 Represent those judgments with meaningful numbers.
4 Use these numbers to calculate the priorities of the elements of the hierarchy.
5 Synthesize these results to determine an overall outcome.
6 Analyze sensitivity to changes in judgment.

This methodology derived from the techniques of analytical hierarchy process (AHP) and pre-emptive goal programming. AHP is a widely adopted decision support technique in management research. For example, the applications of AHP can be found in evaluating risk factors in enterprise resource planning implementation and in translating knowledge of supply chain uncertainty that developed an integrated supplier selection. Moreover multi-echelon distribution inventory model for the original equipment manufacturing company in a built-to-order supply chain environment uses fuzzy AHP and a genetic algorithm. Owing to its advantages of implementing a hybrid method, researchers attempted to address the question of paucity of research with real industrial applications through undertaking a survey on supplier evaluation within a multinational telecommunications company. The proposed supplier selection methodology would indeed assist in reducing the product development timeframe as it automates the evaluation process and provides the procurement team with a flexible and responsive tool for assessing prospective suppliers. The assessment tool includes four types of indices to measure supplier involvement in design, namely satisfaction index, flexibility index, risk index, and confidence index. These indices, nonetheless, measure the extent to which both the customer requirements and the supplier capabilities match or mismatch and therefore reflect the potential or risk of signing a project contract. It may be noted that these indices are limited in measurement nature, and such supplier selection method was not conducted using established quantitative approaches. The AHP helps the analysts to organize the critical aspects of a problem into a hierarchical structure similar to a family tree. By reducing complex decisions to a series of simple comparisons and rankings, then synthesizing the results, the AHP not only helps the analysts to arrive at the best decision, but also provides them with a clear rationale for the choices made. In the AHP approach, the decision problem is structured hierarchically at different levels with each level consisting of a finite number of decision elements. The upper level of the hierarchy represents the overall goal, while the lower level consists of all possible alternatives. One or more intermediate levels embody the decision criteria and sub-criteria. AHP involves structuring a problem from the overall objective e.g. selecting the appropriate six sigma methodology to the sub-objective that minimizes risk, maximizes process capability forming a hierarchy structure. AHP develops priorities on all sub-objectives that is based on predetermining measurements and the decision makers judgments throughout the system that are calculated through pair-wise comparisons. It has been calculated using absolute measurements such as money. Decision makers evaluate each objective against the others within each level. Since its initial development, the AHP has been applied in a wide variety of decision areas, including those related to production and operations management. It makes it possible to incorporate judgments on intangible qualitative criteria alongside tangible quantitative criteria. Elements in each level are compared pair wise with respect to their importance to an element in the next higher level and, starting at the top of the hierarchy and working down, a number of square matrices called preference matrices are created in the process of comparing elements at a given level. These levels represent the decomposition of the overall objective to a set of clusters, sub-clusters, and so on down to the final level. Decomposing the complexity of a problem into different levels or components and synthesizing the relations of the components are the underlying concepts of AHP. There are many factors for application of AHP. The first step in the application of AHP is disintegrating the unstructured decision into components and then arranging them in a hierarchical order. In a typical hierarchy, the top level reflects the overall objective of the decision problem. The elements affecting the decision are called criteria and they are represented at the intermediate levels. Criteria can be subjective or objective depending on the means in evaluating the contribution of the elements below them in the hierarchy. Furthermore, criteria are mutually exclusive and their priority or importance does not depend on the elements below them in the hierarchy. The lowest level comprises the decision options or alternatives. The number of criteria or alternatives should be reasonably small to allow consistent Pair-wise comparisons. The hierarchy does not have to be complete, that is, an element at the intermediate level is not required to function as a criterion for all elements in the lowest level. Thus a hierarchy can be divided into Sub-hierarchies sharing only a common topmost element. Once the hierarchy has been constructed, the decision maker begins the prioritization procedure to determine the relative importance of the elements in each level. Elements in each level are compared pair-wise with respect to their importance to an element in the next higher level and, starting at the top of the hierarchy and working down, a number of square matrices called preference matrices are created in the process of comparing elements at a given level. The decision maker can express his preference between every two elements verbally as equally important (or preferred, or likely), moderately more important, strongly more important, very strongly more important, or extremely more important. AHP was used first to determine the ranking of suppliers, and thereby provide prioritized results of suppliers to aid in model building. Integrating AHP-PGP was secondly presented as an extension to consider additional criteria constraints in the decision making process.

Generally, the correct usage of the developed AHP-PGP approach depends on the following stages: AHP approach calculates the overall scores of alternative suppliers on the basis of the selection criteria. The selection criteria, developed in the AHP model, have been grouped into four main criteria such as manufacturing, technology, business, and service factors. Then, the AHP process has been used to derive weights for each supplier with respect to each selection criteria. After that, the derived weights have been used to establish supplier scores in the utility objective function. Then, all the objectives that are considered important in the selection process are formulated along with constraints. The weights, objectives and constraints are integrated and solved using PGP approach and the results are presented to the user as an outcome of the solution process. The integrated AHP-PGP approach allows the managers to make effective decisions regarding supplier selection. At the same time, this approach permits the decision-team to analyze trade offs in supplier selection decisions while minimizing suppliers' defects rate, rate of late order delivery, purchasing costs, and maximizing suppliers' scores and after sales service levels. Moreover, it also has flexibility to respond to changing needs of the automotive firm due to dynamic decision process. However, the success of the integrated model is based on the accurate measurement of the goals, criteria and alternatives provided by the decision makers. Thus, this model was developed in consultation with the decision makers who were both the users and the managers of the automotive firm. Then, the results obtained from the integrated approach were discussed with them and found to be in accordance with firm's mission, objectives, plans and policies. According to them, the major advantage of this approach is that it assisted them think in a comprehensive way. Similarly, the decision-team agreed that the model would help in examining the strengths and weaknesses of the suppliers by comparing them with respect to appropriate criteria, sub-criteria and alternatives. From a managerial perspective, the integrated model would also be a useful tool to establish long-term relationships with the suppliers. Thus they have accepted to use this approach and to contribute to its efficient implementation. On the other hand, they have suggested that the model should also expand to incorporate criteria such as quantity discount, demand satisfaction and budget constraints for better implementation of the model.

Several software packages now incorporate AHP which fulfill the requirements of a decision support system (DSS). These are techniques that can assist management in the gathering, structuring and interpreting of all relevant information to improve their decision making. Typically, these packages involve: criteria and alternatives specified by the users, i.e. model structures are user driven rather than system driven; weights allocated to each objective or criterion; the user(s) evaluating each alternative against each criterion; a summary calculation that indicates the best alternative based upon the criterion weights and alternative priorities. The usefulness of AHP output is dependent on how well the model represents the problem domain, the way judgments are elicited with different modes of questioning, how adapted ratio scales can capture the true differences between alternatives, avoiding rank reversal, dealing with interaction effects, and applying rating scales for facilitating evaluation. Another major concern is the applicability of the knowledge base for future decision making. When these concerns are satisfied, the user can be more confident and committed with the output. Thus AHP not only looks at manufacturing outlook but calculates all relationship with suppliers.

CHAPTER 13

E-commerce strategies within a business
(BY BAISHAM CHATTERJEE)

Electronic commerce is becoming pervasive and has high impact on business as well as peoples lives. It enhances the competitiveness of organizations by lowering transaction costs and focus on differentiating their products and services. The development of new business models are supported by information technology for electronic commerce. Culture, governance and decision making structures makes the difference. An EC business model is defined as a competition strategy for the marketplace and a structure of business processes for the entire electronic trading like negotiation, purchasing, and logistics of products or service delivery. Customer, electronic intermediary, deliverer and back-end functions are all the important factors for EC. Electronic marketing which is a result of EC applications makes advertising and selling more difficult, but also provides new opportunities for online services and marketing research. Internet based electronic advertisements are low cost, easy to update and convenient for customers to access. Marketing intelligence assists the supplier in creating knowledge about the economy, industry, competitors, pricing, market segmentation and customer purchasing behavior. The EC information system model must describe four components of networked organizational computing on the internet which are unique to conventional information systems. They are system infrastructure, EC business process, knowledge management and web sites.

Overcoming the barriers of distance from using EC will be largely dependent on the detailed characteristics of product/service processes. Closely related to characteristics of product/service processes, customers are likely to prefer services delivered from an approachable point arising from the need for face-to-face interactions, maintenance of service quality, and avoidance of perceived risk in delivery. Meeting customer needs of geographical accessibility is a critical factor in establishing EC operation strategies for customer contact and service delivery, customer support and logistics, and distribution channel strategies and network design. In a very descriptive study, it has been examined that the characteristics of product/service processes and the relationships with customer depend on geographical accessibility. Focusing on the characteristics of product/service processes can provide strategic insights generally adaptable to any products or services in understanding customer needs of geographical accessibility in EC. The differences in the degree of these on-line interactions in EC are closely related to the characteristics of product/service processes as well as business strategies and technology applications. Transactions may require very high or low on-line interaction depending on characteristics of product/service processes, despite high on-line substitution. Many factors affecting customer needs of geographical accessibility are related to the characteristics of product/service processes, similar to the needs of customer support and logistics in EC and the delivery of tangible services. Electronic commerce offers an opportunity to provide service 24 hours a day for those areas of customer service that are suited for digitization. From a customer's point of view, time-to-market is the elapsed time from product idea to deliverable product while delivery time is the elapsed time between order and delivery. Both time-to-market and delivery time for physical products can be reduced by shortening the lag from order entry, and customer information can be improved by opening-up information systems which offers a means of enhancing competition. Electronic commerce customer data increases the possibility to identify and address customer segments as similar as possible within the segment and as different as possible between the segments. Market expansion is either through increased market share or based on the increased span/influence of electronic commerce for geographical expansion into new markets, to achieve economies of scale. The product range can also be extended and used for expansion to achieve economies of scope. Electronic commerce enables business development in activities, processes and services, as well as structural and segmentation change. In addition, electronic commerce development promotes a change of roles and responsibility, within the organization and between organizations. It increases the ability to tailor offerings to the specific needs of each customer while maintaining competitive costs and pricing. Focus extends beyond the focal firm towards the whole supply chain, and on adding value to core products and services. There are two domains for EC applications. The first domain perceives it as an inter-organizational information system (IOIS) which is shared by two or more organizations. It plays the role in inter-organizational communications and business transactions, enhancing the productivity and competitive advantage in this level. Electronic hierarchical coordination deploys information technology to automate or re-engineer the traditional hierarchical co-ordinations, lowering overheads and inventory. The coordination cost through the electronic market is at a medium level when compared to the electronic hierarchy because it is likely that coordination costs will increase in line with the number of suppliers being coordinated. The market cost is higher in traditional markets compared to the electronic market, as more market participation is required in traditional market coordination. The distribution costs are higher for both the traditional and the electronic market, since it is likely that the distribution channel has an ad hoc basis and will, most likely, involve relatively higher distribution costs. The technology cost involved in the electronic market is of a medium level when compared to the traditional market.

One of the ideas in understanding the profit discrepancies in E-commerce are: the industry technology fit theory, which argues that the nature of industry makes e-commerce to be more suitable for certain industries than for others. The industries whose characteristics better fit the e-commerce model are more likely to be benefited, while those having a poor fit will perform worse. The electronic commerce systems are so designed that they act as a central communications base or a central hub. It uses the principle of single source and reduces the number of communication passages. This communication is easier in a secure environment. With the help of the information channel the entire construction supply chain can communicate and archive information throughout the lifecycle of the construction project. Each independent client is provided with a unique project specific web site created around an information database. All data can be transferred to all project team members from the earliest concepts and specifications, through detailed design, building studies, pre-fabrication, construction, maintenance, operation and improvement.

For e-commerce where early in the product life cycle many firms struggle to innovate and gain market share, variation tends to be high which leads to quality advances; later, after industry consolidation take place, variation may decrease which may result in lower quality growth. It also may suggest that industry strategies to keep variation high would benefit customers, although volume shares may not increase for individual firms. Creating incentives to foster greater variation in development may benefit customers in general. The message for e-commerce firms is relatively low innovation rates lead to long-term decline. Firms' ability to copy initial quality advances—characteristic of many e-commerce environments—exacerbates this risk because competitors can innovate themselves and forge ahead by copying any initial lead in an information quality attribute. This illustrates the some advantages of patent protection regarding technological advances. However, patent protection may retard the diffusion of innovation and may result in lower quality levels for the market as a whole. Firms that fall behind in both quality and market share may invest to increase their quality growth rate but depending on how quality relates to market share may gain less ground along the volume dimension. However, in markets where information quality changes do not necessarily translate one-to-one into market share changes for example, where brand loyalty or pricing are factors investments in information quality may not enable firm to recover lost ground.

Adoption of e-commerce has evolved from rudimentary simple Web site construction applications to more sophisticated customer service and personalization models. Initial e-commerce applications by the dotcom companies sought growth as a primary goal while profit considerations were viewed from a longer-term perspective. However, later applications involving brick-and-click companies emphasized competitive advantages and profit considerations and the strategies to achieve these goals. Competitive advantage and profit considerations in the adoption of e-commerce are crucial for small and medium-sized enterprises (SMEs) in that they have limited financial slack to experiment with new approaches and limited cushion for failure. A plethora of failure by the pure dot-com companies with unproven business models have constituted a caveat for the bricks-and-mortar companies to exercise due diligence in venturing into e-commerce. Despite the staggering importance of e-commerce and the strategic considerations for its successful adoption, empirical studies, in this connection, have been scarce in the literature and the research gap is particularly evident for the SMEs, hitherto. The basis of an E-commerce strategy starts with three different levels with the initial level, where focus is mainly on cutting costs and/or raising productivity. At the next level the focus shifts to the use of e-commerce to access new customers and markets. The third level builds upon gains at the previous levels; companies seek sustainable competitive advantage by attempting to achieve a complete integration of e-commerce into the company's overall business strategy. E-commerce strategy focused on customer base expansion is more likely to be adopted by the SMEs pursuing differentiation and hybrid competitive strategies rather than cost leadership strategy. Cost leaders rarely scan the environment for new opportunities, have lower risk-taking propensity and focus mainly on defending their turf. The SMEs pursuing differentiation and hybrid strategies are likely to adopt e-commerce strategy focused on customer service in parallel with the strategy of customer base expansion in order to achieve superior performance outcomes. Although, there is a greater need of online interaction and customization in EC, differences exist in the degree of necessity of online or offline interaction and customization according to the characteristics of the product or service process. Generally, in EC, offline interaction and customization are avoided in order to maintain efficient operations. However, offline interaction and customization are still required in many EC services in accordance to the characteristics of the product or service process. For highly customized services, EC services provide an array of conveniences to customers but the contribution of online channels for market expansion is limited to narrow niche markets with a preference for online communication. Although, EC can generally overcome barriers of distance and lead to dramatic market expansions, compared to the traditional brick and mortar companies, market scope and customer needs of geographical accessibility vary according to service processes including interaction, customization, as well as product characteristics. As such, factors such as services, geographical accessibility and online accessibility are critical for building EC operations strategies. In this study, the focus was specifically on the changed competitive priorities in EC operations strategies. In order to reflect the unique characteristics of EC, competitive priorities linking operations strategy to corporate strategy in EC need to include some additional elements compared to traditional operations strategies. One of the most vivid implications of internet-based EC for SMEs is the potential for external communication and information gathering for market and product research. Although the breadth of activities pursued in EC field is limited at present, the continued growth of EC will enable companies to engage in currently under utilized applications such as job advertisements and video-conference. SMEs need to perceive that benefits of EC will outweigh the costs of EC.

E-commerce generally requires a communication platform that consists of an electronic network for transmitting physical data and terminal devices for users to send, receive and process information. Before an e-commerce has to function and before a transaction has to take place, vendors need to disseminate product information to consumers and during transactions; consumers need to communicate with vendors about orders, accounts and post-sales fulfillment. Communications that require a long time to enable understanding or that cannot overcome different perspectives are lower in richness. In a sense, richness pertains to the learning capacity of a communication. Information and communication technologies have exerted fundamental changes on them tradeoff. Compared to remote selling based on catalog; the internet-based e-commerce enables real-time, two-way information exchange and enhances communication richness dramatically. Also, the open standards of the internet and the wide network access have increased communication reach. As a result, the tradeoff situation has been reduced significantly, leading to the benefits of greater reach and richness at the same time.

Furthermore, we extend the framework by considering costs of communications. As was defined earlier, the major components of a communication platform include a communication network and terminal devices. Accordingly, costs of communications, from the user's perspective, include the costs that consumers bear to access the network and purchase terminal devices. This is important because costs are a significant factor influencing users' acceptance of new ICT. The degree of ICT's diffusion will in turn affect the extent to which the potential benefits of e-commerce can be realized. First, the limitations of mobile handsets in computation power, memory, and screen size restrict the ability of m-commerce to develop one-stop commercial websites. Second, many industries have published e-commerce standards to define product specifications, automate information exchange at the back end of the supply chain, and digitize processes in payment and fund transfer. Such standards make the alliance between vendors offering complementary goods and services possible. However, standards particularly designed for m-commerce, though emerging recently, are limited, especially in vertically structured industrial sectors.

EC permeates all facets of business strategy, enabling organizations to improve business communications as well as business processes. It is also known to benefit all stakeholders in the business such as consumer, supplier, management team and investors. One section of the economy that has clearly benefited from the opportunities that EC offers is the small—and medium-sized enterprises SMEs. EC has helped opened up the market, particularly on the supply—and demand-side for SMEs. Small firms are now able to compete in the same global arena that has previously only been the exclusive territory of multinationals corporations. EC uses Information Systems (IS), Information Technology (IT) or Information and Communication Technologies (ICT) to achieve the provision of information and buying and selling of good and services among business stakeholders. In particular, EC currently makes use of the internet-based technologies because these are the current technologies that provide the widest network of information systems. Historically, the development of an EC strategy might be seen as an element in the IS or IT strategy. However, for many organizations, EC has wider implications than just another IT system. In other words, EC may well be an essential component in the formulation of the overall business strategy. The extension of networks outside the company gave birth to the Inter-organizational System (IOS) and the use of Electronic Data Interchange (EDI). There are considerable differences between EC and EDI; some EC applications may exchange unstructured information, while some others have the need for structured information. EDI is only one facet of an EC implementation. According to, even though EDI is used for regular and standardized transactions between organizations, it is only one part of the overall sphere of EC. The further evolution of the network from an inter-organizational facility to one that reaches out to almost any member of the public set new horizons and EC is one of these. General business risks is a part of EC that has constant threats like fraud, theft, and destruction of corporate assets, along with legal issues such as liability and loss of reputation, are all exacerbated by the open interconnectivity of the Internet. Even though the security issues discussed here are divided into two camps, there are many similar and overlapping areas between the two. Essentially, Internet security concerns boil down to access control, authenticity, confidentiality, data integrity, and denial of service. Transaction security includes the middle three issues plus authorization and non-repudiation.

E-services cape is a modern way of CRM where the development of trusting beliefs is accomplished by the activation of the trust building processes during the different stages of promise fulfillment. When a promise is made within the e-services cape, the intentionality process is initiated to help the customer determine the business motives and intentions, influencing his or her trusting belief in the business benevolence. Enabling the promise invokes the capability process, an assessment of the business ability to realize its promise, which affects the customer's trusting belief in the business competence. Keeping the promise triggers the credibility process by which the customer evaluates the extent to which the business has actually delivered on its promise and develops a trusting belief in the business integrity. The entire interaction with the e-services cape results in the activation of the rest of the trust building processes. Relying on the prediction process the customer makes inferences about the business consistency in delivering the promises it makes, enhancing his or her trusting belief in the business predictability. Finally, with the calculative process, the customer performs a cost/benefit-like analysis of a number of scenarios where the business may act in an untrustworthy manner towards him or her so as to eliminate any such suspicions and increase confidence in his or her beliefs. Trust evolves over time as the customer engages in repeated interactions with promises being fulfilled within the e-services cape. Each time a promise is made, enabled and kept, it is evaluated with the intentionality, the capability and the credibility process confirming the customer's trusting beliefs in the business benevolence, competence and credibility. The level of trust is also related to the experience that the customer gains within the e-services cape. Customers perceive the length of the relational exchange as an investment, which is made by the business and is valued highly enough to deter it from acting opportunistically. The number of business-customer contacts also provides a basis for a thorough interpretation of the business behavior, which enables the customer to predict subsequent interaction.

Electronic commerce is an area of strategic concern to most organizations. Enabled by new, networked technology, e-commerce is rapidly becoming an integral part of the many different ways people and organizations do business and manage information. Through the information technology revolution of recent decades, there has been little understanding of how to measure the real costs and benefits of information systems to the business. This continues to be the case with e-commerce in the 1990s when managers are under pressure to adopt the new communications technology or face being left behind. In the current competitive climate, senior managers seek practical techniques to guide their management of e-commerce and to help them to decide where best to allocate costly company resources. However, many of its key qualities, such as convenience, variety and ease of access to information, are difficult to measure. Much corporate wealth is now being accumulated through nonfinancial means so that it is becoming increasingly important to include intangible assets, often referred to as intellectual, human, social or relational capital, in company reports. Thus E-commerce tends to be the most positive system which can be developed better by designing and making further research by calculating all the steps and advantages of E-commerce.

An E-commerce technology is ubiquitous i.e. it is available everywhere at all times and has global reach and the easiest thing about this is that, it has universal standards of operating. The internet reduces information collection, storage, processing and communication costs while increasing the current accuracy and timeliness of information. E-commerce technology will continue to propagate through all commercial activity, with overall revenues from

E-commerce, the number of products and services sold over the web and the amount of web traffic all rising. E-commerce prices will rise to cover the real costs of doing business on the web. E-commerce margins and profits will rise to levels more typical of all retailers. Traditional well-endowed and experienced big employers will play a growing and more dominant role. Moreover the number of successful pure online companies will continue to decline and successful e-commerce firms will adopt a mixed clicks and bricks strategy. It is important to adopt protocols such as TCP/IP, web servers, HTML and relational databases. It is important to understand and make a clear study of business models and firm and industry value chains and redesign them very often to get a very recent and near future picture. The major business models that are used in the B2B arena are electronic marketplace where suppliers and commercial purchasers can conduct transactions; or may be general or specialized. Another model is the E-distributor that supplies products directly to individual businesses, a channel known as B2B service provider that sells services to other firms. The other channel is known as match maker that links businesses together and charges transaction or usual fees. The last type of B2B business model is known as infomediary that gathers information and sells it to businesses. Another brief abstract of the different types of business models shows that C2C business models connect consumers with other consumers. The marketing functional and the market creator business model is the most important of them. Another E-commerce business model known as the P2P business model enable consumers to share files and services via the web without common servers. A challenge has been finding a revenue model that works. M-commerce definitely talks of mobile communications and is the emerging technology business model of our times. E-commerce enablers' business models provide the infrastructure and strategy for an e-commerce company to grow and prosper. The areas that the internet has affected are related to the industry structure, industry value chain, firm value chain and business strategy that has also affected a change in E-commerce models by affecting cost, substitutes, suppliers, final products and affect a product segment.

The internet is used for developing new technologies like advanced network infrastructure, new networking capabilities, middleware and advanced applications that incorporate audio and video to create new services. More research is being done on optic fiber, wireless web and 3G technologies and the increased bandwidth and expanded connections that help in IP multicasting, which will enable more efficient delivery of data; develop latency solutions such as differentiated quality of service, which assigns levels of priority to packets based on the type of data being transmitted; guaranteed service levels; lower error rates and declining costs. Thus E-commerce research is becoming very important and bringing in new technologies that would make people quicker and give usage of technology more options. Building an E-commerce web site needs 5 major steps like: identifying the specific business objectives for the site and then develop a list of system functionalities and information requirements. It is important to design a system design specification, build the site either by in-house personnel or by outsourcing a part of the responsibility to outside contractors and test the system and implement new ideas in the site. E-commerce suites can save time and money, but customization can significantly drive up costs. Factors to consider when choosing an E-commerce suite include its functionality, support for different business models, visual site management tools and reporting systems, performance and scalability, connectivity to existing business systems, compliance with standards and global and multicultural capability. It is mentioned that there are 6 dimensions of E-commerce security like: integrity that shows information can never be altered; privacy policy which shows that ones personal data cannot be viewed by anyone else and personal information cannot be divulged to other organizations. E-commerce firms must develop a coherent corporate policy that takes into account the nature of the risks, the information assets that need protecting and the procedures and technologies required to address the risk as well as implementation and auditing mechanisms. The other way of minimizing security threats are public laws and active enforcement of cybercrime statutes that are also required to both raise the costs of illegal behavior on the internet and guard against corporate abuse of information.

In order to create competitive advantage in internet marketing the framework is on four dimensions: cost, focus, differentiation and scope. The web leads to frictionless commerce and the end of marketing is based on brands. Internet marketing strategies for market entry for new firms include pure clicks/first mover and mixed clicks and bricks/

alliances; and for existing firms include pure clicks/fast follower and mixed clicks and bricks/ brand extender. The other marketing tools include permission marketing, affiliate marketing, viral marketing and brand leveraging. The other techniques for strengthening customer relationship start being built as a strategy from within a business and include one-to-one marketing; customization and customer co-production; trans-active content and customer service. Online pricing strategies include offering products and services for free; versioning; bundling and dynamic pricing. The different methods of protecting online privacy are legal protections, industry self-regulation, privacy enhancing technology solutions which include secure e-mail and the other types of intellectual property protection are copyright law that protects original form of expression such as writings, drawings, etc. The main e-security laws are as: patent law that grants the owner of a patent an exclusive monopoly to the ideas behind an invention for 20 years and lastly trademark protections that exists in the federal and state levels in North America.

The net marketplace has some key concepts like seller side solutions that are owned by suppliers of goods and are seller biased markets that only display goods from a single seller; moreover buyer side solutions reduce procurement costs for the buyer. Vertical markets provide expertise and products targeted to a specific industry. EDI systems that share various procurement documents like invoices, purchase orders, shipping bills and product stocking numbers serve vertical markets and horizontal markets serve a myriad of different industries like the electronic storefront. The supply chain management system that co-ordinate the key players in the procurement process include: developing long term contract purchase containing pre-specified product quality requirements, whereas supply chain management systems coordinate and link the activities of suppliers, shippers, and order entry systems to automate the order entry process from start to finish. Collaborative commerce brings in supply chain simplification and uses digital technology to share company information. The major net marketplace is E-distributor that is independently owned intermediaries and serves many different industries. E-procurement firms help in long term purchasing agreements and are used to buy indirect goods. Exchanges is another potential net marketplace that connects many buyers to many suppliers in a dynamic real time environment, and lastly industry consortia is also a net marketplace that consists of industry owned vertical markets where long term contractual purchases of direct inputs can be made from a limited set of invited participants. It reduces the supply chain inefficiencies by unifying the supply chain through a common network and the computing platform. Similarly the revenue model in an e-commerce firm is a discussed objective where there are 5 types of models. The marketing revenue model that increases brand loyalty and the advertising revenue model that advertisers are expected to pay through banner and rich media ads. The pay per view revenue model depends on bandwidth capacity that helps view movies, sporting events as content providers. The subscription or membership revenue model that depends on the niche and lastly the value added revenue model that add to the price of a traditional product by charging a fee at the web site for access to premium content.

Thus this chapter provides a clear view of how the internal structure of a business is changing and how the internal structure should be managed in the future with the present developments of the internet and the strategy components that have added to that.

CHAPTER 14

Core competence and organizational learning, understanding the internal quality
(BY BAISHAM CHATTERJEE)

Core competence is like understanding the internal resources of a business and understanding the resources, capabilities and competencies. A clear description would be like: One major benefit for the identification process is that employees survey the organization's different parts, which increases organizational awareness of ongoing and latent activities. A setback in the process is the indistinct use of the associated concepts. For instance, their characteristics are not kept separate, either in conceptual discussions or in practical actions. This is unsatisfactory to those interested in core competence matters, since the characteristics of the associated concepts may enhance our comprehension of the core competence conceptions. Contributions are to the gap in knowledge by investigating those characteristics which discriminate between and signify the associated concepts. In addition, while the identification process and the label associated imply links between the associated concepts and a core competence, these links are still only assumptions in existing research; theoretical and empirical discussions and validations are lacking. Such links, however, are of major importance, since they involve potential influences. The purpose of this chapter is to outline a core competence model by exploring links between core competence and the associated concepts of competencies, capabilities, and resources by proposing refinements to the characteristics of these concepts. The primary signifying characteristic of a competence, apart from its being inherent in individuals and teams, is development. The concept is generally separated into functional competencies and integrative competencies. The former are used in daily activities, and the latter to integrate and develop new competence components. From a technology perspective, scholars suggest that product innovation, facilitated and improved by competencies, are a driving force of firms' renewal. Three types of competencies are distinguished: first-order competencies, which comprise customer and technological competencies; integrative competencies, or the ability to combine first—and second-order competencies, or the ability to build first-order competencies. The typology is based on the same fundamentals as the division into functional versus integrative and exploitation versus exploration, and is relevant to the concerns of this chapter since it studies manufacturing companies with a focus on technology, which is appropriate for the empirical case. Here, we follow the lead, and define a competence as residing in individuals and teams with development as its general characteristic. However, since core competencies are key ingredients in organizational success, they are already highly developed, which implies that minor competence developments are unlikely to have any impact on them. Consequently, only major developments i.e. improvements are included here. Systems play critical roles in many company undertakings, such as structuring core competencies, and routines too are essential to bring order to activities and processes. These critical functions suggest a link of support between a capability and a core competence. Accordingly, capabilities and core competencies are here proposed to be linked by support from routines and systems. Core competencies take part in the value processes by managing the basic elements of a business, occasionally by managing them. In other words, core competencies are utilized in the value process. However, it is not the case that every resource is active in every value process that a core competence does not take part in. Again it implies that a resource is not always linked to a core competence. However, in congruence with the capability concept, it is proposed that a resource is linked to a core competence, since core competencies are utilized in the value process. A core competence is a competence fulfilling three criteria. This implies that competence, indicated by improvements, is linked to a core competence. From the theoretical discussion, we can also propose that a capability, indicated by the support it supplies, is continuously linked to a core competence. Resources, however, seem to not always be linked to a core competence. Here, it is proposed that a core competence is generally utilized in the value process, that is, by resources, and so, accordingly, they are linked.

Skill is a resource difficult to indicate and measure due to its intangible character. Yet, despite its intangibility, this particular manufacturing skill is highly regarded and was declared by customers and employees to be of major importance for the development and success. The company has built excess capacity into the manufacturing processes, due to the importance of delivery-safety for some customers with high time—and delivery-sensitive businesses. Transformation has put a heavy pressure on the competencies of the sales unit. However, the newly acquired and developed internal system has supported the transformation. A manager described a system for current and future business, one which is directed towards the satisfaction of customer needs. The system makes it easier for employers in sales and marketing departments to manage the ways in which communication with the customers is maintained and developed, by sharing information across divisions and departments, by managing major customer projects, and by keeping track of the customer's product portfolio. It is emphasized that the system is crucial to customer loyalty, providing that two conditions are satisfied. Firstly, it is vital that the system is accurately maintained with relevant and up-to-date customer information. Secondly, the system must be easy to use. If information systems are difficult to use, or if they contain redundant information, then employees become reluctant to use them, and may even ignore them altogether.

Existing competence theory focuses merely on development characteristics: functional competencies belonging to markets and technologies, and integrative competencies. Since, almost all empirical competencies in the case, company could be categorized as functional and integrative competencies; there was a clear need for refinement. Adaptation is proposed as a new characteristic; it explains how it is possible to initiate, maintain, and develop customer loyalty. A second proposed new characteristic is transfer, by which is meant internal transfer of competencies, for instance between divisions. The capability concept is linked to the core competence concept by support from systems and routines, a linkage which was also emphasized by the empirical case. The assumed capability characteristic of systems and routines was insufficient to describe their support for the empirical core competence. A capacity characteristic is therefore proposed, along with a communication characteristic. The former refines the capability concept by adding accuracy; the latter refines the same by adding an active direction for the support. Finally, it was not possible to refine the resource concept in itself; all we were able to add was the proposed intermittent link to core competence. If information systems are difficult to use, or if they contain redundant information, then employees become reluctant to use them, and may even ignore them altogether. If this happens, it is possible that the support function of a capability may become unproductive and eventually obsolete; both imply that a system will not be used by employees. This suggests that the concept of organizational capabilities should be supplemented by a characteristic concerned with accuracy: measuring to what degree the support is redundant. The concept described here as accuracy shares its basic notion with the capacity characteristic of the capability concept.

Core competence is the ability to operate efficiently within the business environment and also the ability to respond to challenges. In order to respond to challenges surrounding their business, managers are dedicating to build their core competencies. Core competencies are the collective learning which guides business owners and workers in how to best coordinate diverse production skills and integrate multiple streams of technologies. Furthermore, it is the ability to communicate and be involved in and committed to working across organizational boundaries. It involves many levels of people and the entire array of business functions. Core-competencies have created high tech firms and traditional manufacturers that differ from each other in terms of the business environment. In contrast to traditional manufacturers, high-tech firms often have the characteristics of high-speed innovation, a short product life cycle, rapidly changing environment, and are highly dependent on advanced technologies. Further, the weights of environmental elements are different widely from each other. Thus, it is interesting to explore the core competence both in the high-tech firms and the traditional manufacturers as these two have distinct business environments. Strategic planning in high-tech firms is also regarded as an important dimension in constructing core competence. Environmental sensitivity and speedy and appropriate reactions are relatively more important capabilities for high-tech firms looking to create core competence. Finally, the capabilities of supply chain management and logistics management for traditional manufacturers significantly affect core competence, because these businesses must focus more on services. Thus, these two management capabilities are correlated with service; supply chain management can strengthen the linkages with partners to promote service capability with the purpose of logistics management being to coordinate delivery and shorten the lead-time of delivery.

The competence led perspective differs from the position approach by tending to work from a different starting point and laying its strategic bets in a different place. The positioning perspective starts by considering what industry we want to be in and the position we want to take in it. The competence approach on the other hand begins by assessing which distinctive competences we want to build, and then considers the market opportunities that would exploit those best. A core competence typically has a technological dimension, but it also has governance and collective learning dimensions that integrate into the social fabric of the organization. It is this multi-dimensional and organizational character that makes it difficult to replicate. The second idea that has three levels is the relationship between core competence, core product and end product. The core competence perspective offers more help to strategists pursuing corporate development through organic growth. To begin with it differs from the positioning perspective by stressing that effective business strategies can also be developed from a coherent corporate strategy. The biggest implication of the core competence logic is that it expands considerably the degree of product market diversity within which related diversification can be pursued and generates a wider range of options for organic growth than a product market focus alone would. It appears that core competence is where each organization has increasingly to prove its added value. This leads to a higher pressure of work in the organizations, as proving the added value places a lot of demands on innovation, creativity, customer friendliness and quality improvement. The battle against the competitors can no longer be conducted on the basis of a reactive competitive policy. It has to be conducted using a proactive policy, based on the inner strengths of the organization itself. Business success is increasingly based on the improvement and strengthening of the core business of the individual organization. It is not about always being just one step ahead of the competition but about being continuously ahead of the competition, by mobilizing individual strengths, knowledge and expertise in the organization. In short, this means appealing to and utilizing potential in the organization as the success factor for an organization's survival. Strategic intent, on the other hand, can be likened to an ambition, a mentality, wanting to be the best. It is an entrepreneurial mentality with a specific will to survive and win. Strategic intent involves the definition of an ambitious future and the will to bridge the gap between the present situation and the view of the future.

Core competence can be defined in many ways; moreover another way of defining it is that it is a unique combination of business specialism and human skills that give expression to the organization's typical character. Core competences are the company's characteristic areas of expertise and consist of the synergy of resources such as motivation, employee effort, technological and professional expertise, and ideas about collaboration and management. Working systematically and structurally with core competences gives the organization considerable strategic power. Core competences are difficult for competitors to duplicate because they are distinctive and specific to each individual organization. Concentrating on the core competences makes the organization very effective and therefore results in a competitive advantage. Core competences appear to be the success factors of excellent organizations and trend-setting companies. Core competences are the result of a joint learning process in the organization and they form products in which internal and external business strategies, production logistics and individual competences find their expression. Core competences are therefore supremely suited to enable organizations to respond to their environments and to follow a specific product/market policy. Individual competences are also very important: The content and form of the core competences are created through the connections between the organization's objectives, strategy, structure and culture, as well as its management concepts, the expertise of its employees and the degree to which the employees are appreciated by the management. The skills and motivation of the employees are important strategic aids in the realization of the company's objectives. To derive the maximum benefit from an organization's core competences, it is extremely important not only to recognizing the expertise and skills of employees but also to pay attention to the underlying motives and qualities of the employees. These underlying motives are also known as individual competences, in contrast to the organization's competences. Individual competences are concerned with the fundamental personality characteristics that are inherent in a person's actions in relation to all kinds of tasks and situations. Competences are more important than are knowledge and skills for the successful performance of complex professional management tasks with a high level of responsibility. As far as knowledge and skills are concerned, many people are equal, as can be seen from, for instance, diplomas, certificates, and work experience and work results. It is precisely the employee's effort, enthusiasm, motivation and underlying self-image that distinguish the successful employee superior performer from the unsuccessful one. The two case descriptions at the end of this article examine in more detail the identification, appraisal, remuneration and training of individual competences. The core competences of the organizations concerned form the basis for the HRM set of instruments that are demonstrated.

The compelling reason for switching to a competence-based organizational type is the improvement of the organization's level of performance and the creation of a competitive advantage. The capacity and motivation of the individual employees are essential in achieving an improved level of performance, provided, that is, the capacity and motivation offer support to the organization's strategic line. In competence-based organizations, it is necessary not only to pay attention to identifying and developing individual competences, but also to take personal competences as well as the organization's core competences as the starting-point in recruitment, selection, appraisal, remuneration and the career policy. This total approach is attractive to employees who are creative and innovative, want to learn new skills, to have more responsibility and to be continuously adding to their own expertise. Here, too, it is the professional vocational groups that find this type of organization most attractive. However, it has to be a challenge to each employee to be called on for his or her individual skills and then to be appreciated and rewarded for them. This, increases involvement and motivation which, in turn, increases the organization's ability to increase performance levels. However, the problem with this total approach lies in finding the right match between, on the one hand, the available competences of the employees and, on the other hand, the objectives, strategy and core competences of the organization.

There is a disturbing tendency today to identify characteristics such as quality products and a good reputation as core competencies, when these characteristics are really the result of performing discrete activities well. In failing to associate specific, underlying activities with these claimed competencies, managers are unable to focus on preserving and strengthening the building blocks that create quality products in the first place. Business analysts must dig deeper. When they do, they will find that real core competencies are tangible value added activities that are performed more effectively and at lower cost than that of the competition. These unique and enduring activities constitute a firm's core competencies. Sometimes sharing and nurturing core competencies within a corporation is not enough. As markets evolve, new activities may be required. Moreover, in today's global marketplace, even giant corporations blanch at the cost of launching new products and entering new markets. With product lifecycles shrinking and R&D costs skyrocketing, some companies find it easier to embrace their competitors rather than fight them.

Competences that consider the ability of an industrial enterprise to deliver customer orders at the promised day and with the specified quality and customization is at the heart of the competitive strength of the company it is complex, more difficult to imitate and less dependent on technology more dependent on knowledge. However, it is difficult to identify the third type of competence. Examples of some of the elements of complex competences are the: quality management system; production management system; tacit knowledge of individual employees interacting collectively; and attitude and organizational culture of the company. Examples of complex competences could be the entire process of product development from idea to actual manufacturing or the process of handling incoming orders in a company. It is common for such complex competences that they tend to be more like business processes stretching across departmental and other barriers and a lot less like clearly delimited and structured technology like competences. Complex competences may include a number of competences which are complementary. Hence, the resulting competence is greater than the sum of the individual competences. It can be found that businesses following the SBU techniques were often dependent on external sources for critical components, such as motors or compressors, which would contribute to the competitiveness of a wide range of end products. Unfortunately these critical components, the physical embodiments of core competences, were in the direct control of an outside agency all as a relationship of core competence. It is generally believed that the most important resources and capabilities are those which are most difficult for competitors to imitate and difficult to understand, provide potential access to a wide variety of markets, make a significant contribution to the perceived customer benefits of the end product, and very durable. Much of the confusion within the different management circles arises from straying from these constraints. A long list of competences is an indicator that at least one significant constraint is missing from the working definition. Some examples of core capabilities would include innovation, short product development cycles, well-motivated employees, and a strong service reputation. Core competences are very dynamic. Since they arise from the talents and skills of individuals, and are the result of teams and organizations collectively. They can also be easily lost or changed. Key competences can be subject to the negative influences of company policies, internal politics, and other direct influences. In addition, they can leak out of the company by the injudicious use of outsourcing, or through alliances with other companies. Some companies worry that a loss of specific employees would be a loss of a core competence. This should not be the case for a competence correctly identified, since one of the restraints is that it is difficult for a competitor to acquire. Moreover, if a competitor were to acquire some key personnel, these individuals would no longer be working under the same conditions which supported them earlier. A core competence is a part of the working system, which includes the employees. If a significant portion of the system becomes outsourced, downsized, or laid-off, it is indeed true that a core competence may be adversely affected. Core competence activities are still viewed by many as an expense, rather than a business development investment. This has stalled the adoption of core competences in some organizations. A firm which adopts core competences would have a strategic platform from which to work that would allow easy movement to take advantage of a tactical opportunity. Current competence proposals are beginning to emphasize the employee rather than dwelling on technological capabilities alone. Some believe that the organizational capabilities are best improved by developing the expertise of employees. Enhancing organizational capability can be used to develop new products and services, and bring them to market more effectively. This technique helps to support the concept of complementing front-line entrepreneurship by emphasizing distributed, rather than core competence. In supporting this, top management would leave the operating units with the task of creating the competences needed to pursue local opportunities, while ensuring that the new developing competences are shared throughout the remainder of the corporation. With the dynamics required of current companies, and with competitors trying new concepts which are very different from the business norm, organizations which can learn and grow to be more flexible will have a tremendous advantage. However, flexibility requires resources to be available for execution when the opportunity arises. Otherwise, flexibility and all other organizational efforts will have little value. Learning the competences or capabilities that will be called on will be a critical long-term investment.

The process of identifying core competencies usually entails having employees identify core competencies by scanning and assessing company critical resources, capabilities, and competencies three factors commonly referred to as associated concepts. In the identification process these concepts often become conceptually and empirically merged something that occurs in strategic management research too, when these associated concepts are defined interchangeably. Although merging the associated concepts is occasionally justifiable, it normally makes sense to distinguish them by their characteristics. In fact, each concept is acknowledged to be substantial enough to have its own major research stream in the strategic management field, namely, the resource-based, competence-based, and dynamic capability-based streams. Although neglecting the associated concepts distinguishing characteristics may occasionally be useful in complex identification processes, for more advanced core competence matters doing so are unsatisfactory. The very diversity of the concepts enhances our understanding of core competence, and is relevant to research issues such as core competence management, a matter going far beyond mere identification. The competencies hierarchy represents a combination of two types of hierarchy: a cumulative one, since the concepts are described as building on each other, and a qualitative one, since higher-order concepts are assumed to be of greater organizational value than lower-order concepts are. The hierarchy starts with resources at the bottom; next come capabilities, which build on resources, and then competencies, which build on both resources and capabilities. Finally, core competencies reside on the highest level; they are of the greatest value to organizations, but are the most difficult to achieve. This hierarchy notion seems reasonable, although the definitions of the associated concepts are not explicitly presented. In fact, the hierarchy suggested seems to be valid only for structuring the concepts, for revealing that they have different organizational applications. In other words, the concepts' differing characteristics as such are neither acknowledged in the hierarchy nor applied in practical matters. The capacity concept is too vague and difficult to identify or measure empirically, so it is disregarded when considering the definitional issues. Systems play crucial roles in many company undertakings, such as structuring core competencies, and routines are also essential in organizing activities and processes. These crucial functions suggest the existence of a supporting link between a capability and a core competence. An empirical study recently conducted by the author found evidence for the existence of the supporting link, since empirical capabilities involved both internal systems such as ISO certification in quality and environmental issues, and decentralized efficiency in administrative routines. The main characteristic of the empirical capabilities identified was that they supported organizational activities and processes. Accordingly, capabilities and core competencies are here proposed to be linked by support provided by routines and systems.

The concept of strategy competence comprises an international market experience dimension. High familiarity with international market conditions particularly strengthens the competence and the possibility of finding sustainable positions in local markets. Local market conditions may be specified by aspects such as market entry barriers, the behavior of competitors in the market, consumer behavior in the market, and local preferences for technology and choice of suppliers. Familiarity with international market conditions might be acquired through any sequence of market entry or any way of searching knowledge on the conditions external to the firm. Besides international market experience, the overall strategy competence for a particular firm is due to the position of its businesses in relation to the core business of the corporation. This is manifested in the business relatedness dimension of the concept. It means, for example, that high relatedness enables accumulation of related and relatively homogeneous knowledge on products and markets, and that efforts to establish a business that belongs to the corporate core business hypothetically possess strong potential for success. The opposite is valid for a unit less related to the core. Here, core competences are meant to be: the collective learning within the organization, especially how to coordinate diverse production skills and integrate multiple streams of technologies; the organization of work and the delivery of value to the customers; and a guide to patterns of diversification and market entry. In general, at least three tests can be applied to identify core competences of a company. First, a core competence provides potential access to a wide variety of markets. Second, a core competence should make a significant contribution to the perceived customer benefits of the end product. Finally, a core competence is the kind of competence that is difficult for competitors to imitate. The competence to innovate in a particular domain follows consistent investments to develop the facilities, personnel, intellectual property, inter-organizational relations, and tacit knowledge to successfully innovate in that product or technological area. This means that the knowledge stock a firm has accumulated in a technological subfield conditions its returns to research and development investments in that subfield. Therefore, it is natural to expect that strategy development related to the core business will produce superior results when research, development, and market investments are concentrated in areas of a firm's established competencies. Building strategy competence, thus, requires a focus on managerial relatedness and a drive towards exploiting management skills and leveraging routines that are applied in the core business. The chances of reaching high performance in the local business will also increase if the need for investments in market communication is taken seriously, and where there is an explicit ambition to achieve similarity in brand recognition through efficient brand management. This implies that the international firm has to evaluate the difficult question of balancing standardization and adaptation of communication messages between different markets in relation to the internal process of brand building, covering links among core competence, core products, end products, core values and brand identity. Anyway, an efficient branding process will facilitate the break through of existing customer loyalties and the search for appropriate distribution channels.

The organization can face its crisis and choose to move into introducing New Market Values by deliberately choosing to innovate through doing something different which initially is high-risk, with potentially high reward. In time, this high-risk and reward strategy stabilizes by having introduced a market standard which can become a low risk and has the potential for increasing as opposed to diminishing returns. At which point, competitors enter and copy, and the cycle begins again. The basis of this kind of thinking, knowledge leadership is illustrated within a variation of the Johari window. In this modified Johari window, knowledge leadership exists in terms of that which has a potentially high strategic impact and is known only to the organization. What is public domain or known to others but has high strategic impact requires co-operation, if only to manage the timing of its impact; and that which is of low strategic impact and only known to you can be shared or exchanged possibly with academics. Finding a good fit between the firm competencies and its environment is the core of a good strategy and calls for the optimization of both external effectiveness and internal efficiency. Acknowledging that both factors form the underlying incentive to the fit, relevant adaptive processes and an ability to change one's business scope is necessary due to the environmental change. The very basic need of how a firm can survive in the long run implies that managing the environmental change strategy making at the firm level becomes inevitably a continuous management exercise. During it the firm management has to face and find an acceptable and workable answer to a couple of the most fundamental questions: where to go and how to get there. Hence, when properly managed a good strategy defines the firm's future direction and gives guidance as to the operationalization of management activities and the use of resources. Given that firms are heterogeneous by their resource base, the emerging resource-based view of the firm indicated that firms can achieve good performance through their different skills, capabilities and firm specific use of resources. The inherent and rapid development of ICT technologies in particular proves that a number of new ways to compete in the current environment can emerge. Among other things the rapid evolution of new production methods has made it possible to produce individually varied products at the cost of mass products. Also time competition has turned out to be a prominent field of competition which the conventional strategic thinking has not taken into account in any specific way. The sources of competitive advantage are superior skills and resources which in turn make it possible to achieve positional advantages through superior customer value and lower relative costs. This also corresponds to the prevailing resource-based view of the firm indicating that intangible resources in particular are capable of creating sustenance. As to organized competencies, learning is the glue by which change and development of competencies can be linked. Firms improve their strategic flexibility for the sake of competitiveness which is why operations in a highly dynamic environment may demand that a considerable part of the firm activity must be directed to competence building. Given that competence building implies complex patterns of coordination and cooperation between the employed resources, successful competence building also explains the emergence of distinctive competences and the capability to create corresponding customer value. Acknowledging that capabilities are what a firm can do and that they are subject to learning from repetition, skills and accumulated knowledge form the resource base of a given capability. As a consequence, the knowledge which firms may possess or achieve forms ultimately the foundation of competencies and raises the question of how to acquire new and useful knowledge and manage other related assets simultaneously. A qualitative change of assets, where competence building by definition implies, can take place by creating and adopting new capabilities which in turn enable future business options. It is more specifically stated that continuous learning from the markets becomes an extremely important matter calling for such qualities as open-mindedness, scanning ability or continuous experimentation. A given reorientation capability together with marketing development forms the basis of successful implementation. Independent of the final business scope to which identified ideas may give any reason, a number of change-generating capabilities and competences must exist, which is why learning and competence building in their different forms becomes topical and an essential ingredient in the business concept construction. Depending on the targeted strategy the key dimensions of value generation are either costs or differentiation. As to the costs, better value can be generated through effective control of costs or by the renewal of the whole value chain. However, as to the appropriateness of the value chain as mentioned before, also the role and positional status of assets has changed so much since the emergence of the concept of the value chain that traditional chain-like patterns are not workable anymore. This also applies to product differentiation, the scope of which has broadened to consist of the whole product offering instead of single products only.

A sense is capability: competences being the kinds of skills and knowledge that an individual or organization possesses in a specified area of performance. It is apparent from the literature that the differences between sufficiency and capability competence are not clearly understood. It is, for example, quite possible that an organization or individual could be classified as highly competent in a capability sense, having a well-developed set of competences that can be classified as incompetent in the sufficiency sense if the business tasks performed require proficiency in competences. The confusion has been added to by the use of minor variants of language: competence competency; competences competences. Another competence element is the company's market orientation. A decisive success criterion for the company is that its products provide value to consumers. Therefore, it is very important that the company involves consumers from the very beginning of new projects preferably basing idea generation on consumer demands and needs. The third competence element that directly supports the product development competence is the company's ability to follow the clear objectives set for its product development both short—and long-term objectives. To a large extent this competence element rests on personal characteristics of the product manager who typically also is the project leader. This competence element, know-how and market knowledge is dominated by two other blocks of competence elements. The first block is related to the company's ability to analyze the market. Looking at that block of competence elements one sees that the competence element is supported by, among other elements, its ability to buy external market analyses. As the company does not have specialists that can conduct all the necessary market analyses, the company's ability to buy market analyses becomes crucial in connection with product development.

While the generic competencies approach has a legitimate part to play in the reconstructing of organizational competence, we would assert that it is only a part. Reconstructing organizational competence and what makes for competence of performance is a bigger issue than simply identifying generic lists of competencies assert, and one which many current competence models largely fail to address. The volatility of the business environment demands that organizations are continually examining their competence. Thus, organizations will continually need to change, to adapt, and to better position themselves. A fixed list of generic competencies sits uncomfortably with this view of organizational life one in which organizations may well need to continually reinvent themselves and in which competence is continually being reconstructed. Many writers point to the need for different lists of competencies based on different organizational performance criteria and different business scenario. We believe that fixed lists of generic competencies may create the belief in management that, at least, that part of the puzzle is solved; a belief which the authors suggest may be, at its best, illusory and, at its worst, disastrous. It is a natural belief that many competency-based approaches to addressing issues of organizational competence are still too narrowly focused, too static in nature and more tactical than strategic. Beyond fixed lists of competencies, senior managers need a model which caters for the dynamism of the worlds that they and their organizations inhabit. Our research set out to construct a more comprehensive framework and methodology to help senior managers explore issues of organizational competence and to address requirements for change from a strategic perspective. A further influence on the definition of competence must come from the environmental conditions within which the organization is operating e.g. competitor activity, changing market requirements, economic factors. From a consideration of these two key influencers, competence was defined as: a statement of value accorded to another person in a particular culture and within a particular business environment. Arising from this, it is possible that different organizations can define competence in different terms, depending on their particular organizational cultures and the environmental conditions which they face. Competencies are themselves built up of smaller units termed elements of competencies skills, knowledge, characteristics, traits and abilities. The composition of these units and their relative influence within each individual competency may give some clues to issues of style within an organizational context. The elements of competencies are defined simply as: constituents of individual competencies. Having accepted the importance of organizational culture in combination with organizational environment as determinants of organizational competence, they better understood the need to deal with aspects of culture change. Furthermore, they understood that redefinition of organizational competence would lead to the determination of role composition for partners and the identification of relevant competencies. In summary the framework helped to focus attention and gain a clearer grasp of the complex issues involved in promoting the development of real competence within the firm.

Manufacturing competence is therefore represented by the degree of consistency between the importance given to a capability and the firm's strength with regard to that particular capability. One can conceive that a firm's manufacturing function is competent thus contributing more to the realization of business strategy if it has strong capability on a variable such as flexibility that is also considered important. A firm's manufacturing function is not competent if it lacks competitive capability on a variable which is considered important. On the other hand, if a firm is strong on what is considered not very important, one could argue that this particular strength does not add too much to the firm's manufacturing competence. There are many ways of deriving manufacturing competence and they can be defined in two ways:

First, what are the measures that represent relevant business performance. Unlike other traditional research areas in operations management i.e. inventory control or scheduling, the field of manufacturing strategy has not yet developed a well articulated set of performance measures with which to test a theory. Since issues in this field tend to be broader in scope, a general tendency is to use a set of business-level more specifically, financial and market performance measures, such as profitability, market share, or growth rate. We do not intend to challenge this general tendency, nor is it our purpose here to discuss the single most important performance measure to be used in manufacturing strategy research. Our study is more focused on exploring possible linkages between business performance measures and manufacturing competence. Second the model of manufacturing competence starts with the assumption that there exists a sound business strategy. It is not addressed how a business strategy should be developed within a particular competitive environment, nor is it tested whether a specific type of business strategy affects various performance measures differently. A part of the business strategy could well be to determine what performance measures either return on assets, growth rate, or market share should be emphasized. Manufacturing competence as defined here may affect business performance differently depending on the nature of business strategy. Future studies should investigate questions like, if a firm's business strategy calls for a dramatic improvement in market share, then does the manufacturing competence matter, we classified these issues as strategic competence. The average US manufacturer perceives that the capabilities to provide conformance quality, on-time delivery, and reliable products are most important in order to compete in the market. Following these three areas are performance quality, low price and fast delivery. Among the least emphasized capabilities are rapid volume and design changes, broad product line and rapid mix changes. It is clear that such narrowly defined competences do not represent the full scope of management, and this gives rise to the question as to whether such a process of evaluation is in the best interests of managers, and in the practice of management education. If the claims of advocates of managerial competences are correct, then management education can look forward to a golden age. Certain and clear identification of the appropriate content of courses at all levels becomes a practical actuality; and students who seek to become effective managers have a straight and easy road to follow. The likely response to such a suggestion is that competences, as currently defined, are not yet sufficiently sensitive or sophisticated to enable the accurate assessment of managerial performance. A point to be considered is that the competence view is concerned to provide accreditation of experience and is not able to evaluate expertise. An effective demonstration of a managerial skill does not simultaneously demonstrate that the manager has the necessary expertise to judge when and if the use of that competence or skill is appropriate in another situation.

Core competency is rooted from individual level and that is well connected to organizational competitiveness, shifting academic attention of core competency from micro—to macro-perspective. The concept of core competence is derived from the competency that highlights a close linkage to the strategic thinking; therefore, the concept of core competence even though originated from individual level can be easily linked to organizational level. From then on, strategy management combined the concept of core competence with resource-base essence into the strategic thinking and implementation process. Core competency model needs to link the competency in individual as well as in organizational levels so as to successfully formulate strategy. Therefore, it is important for a company to have a systematic and comprehensive framework to accommodate both perspectives in strategy formulation. A clear mapping of strategic content and context would facilitate a company to attain its strategic goals and intention. Establishing core competency is useful for a company to accumulate its resource-base value and to enhance the implementation of the chosen strategies. After a long-term observation from industry it is illustrated that a firm's competitiveness can be strengthened only when its strategies are delicately formulated and its human resources are qualified with solid core competencies.

Core competency has a strong relationship with **organizational learning** and both are dependant on each other in this era of knowledge management. Organizational learning is activity or processes of learning in organizations and exists without any efforts. But it is true that core competence exists in all high tech organizations, but further research can make things more clear and develop on that. Organizational learning is of two types: Adaptive learning or single-loop learning is the more basic form of learning and occurs within a set of recognized and unrecognized constraints that reflect the organization's assumptions about its environment and itself. The constraints limit organizational learning to the adaptive variety, which usually is sequential, incremental, and focused on issues and opportunities that are within the traditional scope of the organization's activities. Generative learning or double-loop learning occurs when the organization is willing to question long-held assumptions about its mission and capabilities, and it requires the development of new ways of looking at the world based on an understanding of the systems and relationships that link key issues and events. It is argued that generative learning is frame-breaking and more likely to lead to competitive advantage than adaptive learning. The difference between organizational learning and learning organization and the presence of core competence can be developed through analysis and futurist acts. High-tech businesses are becoming a present scenario in modern business and in any business core competence is always present and it has to be understood and developed on. Whenever a problematic situation arises in an organization, employees enquire into it and takes the role of organizational learning ex-core competence is getting difficult to manage with complexity of innovation and thus we have to take into view the role of organizational innovation. The focus on innovativeness is one of the two strongest practical reasons to go through the pain of creating a learning organization from an existing organization. Organizations clearly need to become more innovative as we approach the millennium; they should have been more innovative all along: innovative in terms of new products and services that they offered, innovative in terms of new methods of organizing, innovative in terms of how they dealt with their employees, more innovative in general.

The main distinction concerns what constitutes learning, separating behaviors that concern the ability to make incremental adjustment as a result of environmental changes, goal structure change, or other change, as adaptive behaviors that do not constitute learning. It is the development of insights, knowledge, and associations between past action, the effectiveness of those actions, and future actions that constitute learning. Learning is then further divided along low-level learning and high-level learning. Low-level learning refers to behaviors that may, or may not be repetitious of past behavior, often at a routine level, but that form cognitive associations. Transformational learning implies significant change, and change occurs only in reference to an earlier state, as a social construction. On the other hand, incremental change is less perceptible; it does not necessarily result in a recognized change.

For a business organization to develop, it has to be: visionary; developmental; active in learning; and practical. And to shape these four characteristics, the organization has to build management systems for creative environment, performance, knowledge, and concept development. The building of such systems means: accessibility and convenience for knowledge sharing through development process creative environment management; opportunities for R&D personnel to exert autonomy and learn actively self-illumination in management by objectives performance management; instructions to enhance employees' ability in pursuing the leading status in technology and products ability development; wisdom sharing among all members in deciding the focus R&D strategy this is also called concept development management. To develop the internal self development performance and communications, it is important that—in particular, the communications of organizational policy statements. An example was cited where an organization strategy concerned with the marketing function had failed, with a drain in resources. Poor communications result in failure of strategy, similarly sales and marketing or a sales force cannot exist with poor communications. Barriers to learning are here considered as employees' dilemmas that do not contribute to organizational learning. A dilemma is a situation where the individual is supposed to make a decision and act, but there is no obvious decision alternative better than another. Therefore the dilemma demands reflection and reconsideration of the situation to make the best decision. A situation being a dilemma often leads to organizational ineffectiveness if the individual is forced to act. A dilemma also leads to ineffectiveness if the individual finds it easier not to act, when action is needed. A problem is a situation where there is a rational solution which the individual is able to solve. An organizational dilemma therefore demands some kind of organizational learning, and if this learning is not at hand it is interesting to inquire why. These reasons are here called organizational barriers. There are several reasons for barriers: metaphors carried by employees in the organization are not understood or considered by management; visions carried by the employees concerning the future of the organization and concerning the individuals themselves are not consistent with visions carried by management; power differences and communication structures are not changed to support the new ideology; management styles are not changed according to the change in management ideology; official and unofficial norms for behavior mirror past ideology, not the ideology of organizational learning. The Taylor inspired organization of work was replaced by an organization that demanded each person to understand, take part and control the entire work-flow, from raw materials to customer needs. Foremen were replaced by group leaders in each product-flow work group. Managers now function as service units to the product workshop and to customers' wishes. The structure of the organization takes the form of a matrix organization. The new organization meant organizing into product shops, introducing flow-groups, changing work-flow layout, and reviewing the administration and control system. This new work organization implies a new way of working and thinking for employees, managers, clerks and operators as well. Work is organized in work flow-groups, with different goals ahead. These new values guide work and are revolutionary, compared to earlier management norms, when Taylorism was still used in the workshop. Many managers and employees have difficulties in changing their attitudes, communication and actions according to the new situation. If management also focuses on cooperation in flow-groups as concrete goals for action vision, operators and other employees would be able to start acting accordingly. In fact this is something that on a small scale has been done in the studied firm, by building meeting places in the product shop where operators get information via white boards about capacity, productivity and statistics. The operators' view about these places is that they have improved their ability to discuss problems in the work process. It is disappointing that mostly these are not places where white-collar workers and operators share information inquiring into problems together. Another dilemma is that all employees do not want to take part in everyday decisions about work. Workshop operators do not want to acquire more responsibility in some instances in their work situation. They are satisfied with the fact that their workshop manager takes responsibility for their work. As mentioned before, a norm that has spread among some operators is not to participate in project groups. This could be found among all operators, but especially among elderly workers. These operators find it more in line with their competence to do a job, whilst not embracing working in a project group. Writing, discussing and seeking solutions to problems in the workshop which are not obvious are not what this category of operators call doing a job. Mental models and visions in the workplace differ substantially in these cases between operators and managers. A short-sighted view about productivity in the production process is management's norm for the firm, and thus also for employees. All these short-sighted actions hinder long-term planning for a learning organization. Shortsighted planning is institutionalized in the organization structure by planning procedures such as eight-week capacity planning. Another aspect of barriers to learning in the organization structure is the division of work between, for example, the personnel manager and the finance manager. Tradition carries certain norms that specify what should be done by each one in the hierarchy. For example, the chief of economy is supposed to collect quantitative data, while the chief of personnel is supposed to collect qualitative data.

Organizational learning deals with an emphasis on broad and diverse participation and interaction, as well as constant interactive communication throughout the entire organization. This characteristic can be supported through approaches such as meeting times and facilities to facilitate broad interaction and creating sharing events and programs designed to stimulate diverse interest and involvement. A number of technology tools exist to provide the foundation for flexible, productive, collaborative teams unconstrained by geography, time and organizational boundaries. Incentives and reinforcements provide pressure away from equilibrium. Pressures and incentives may, in some cases, serve as disturbances that help provoke the emergent organization. Pressure to learn may come from group norms and from simple awareness of the skills acquired by others, their contributions to success and the opportunities to learn new technologies and skills. Pressures to change may come from learning, incentives may include reward for skills acquired or the resultant contributions, as well as the incentive of a visible selection process providing greater organizational opportunities. Incremental improvements in organizational performance occur when the rate of organizational change exceeds the rate of environmental change. However, fundamental performance improvement can occur only when the rate of organizational learning exceeds the rate of organizational change and, in addition, the rate of organizational change exceeds the rate of environmental change.

The practical consequences of a self-organized, tacit learning system include: persistence of barriers to learning which can not be identified nor removed; existence of walk-talk gaps which cannot be recognized nor addressed; and generation of erroneous conclusions based on non-systemic interpretations of organizational experience which cannot be tested nor corrected. Few steps of the organizational learning cycle are in 4 steps: observe, re-model, interpret and enact. The first step is rooted in an active process of reflective inquiry into an aspect of the organization or its relationship to the environment. The particular aspect at hand presents the focus for inquiry and reflection. This step is primarily self-observation, because the learning unit will always have a rich system of interrelationships between themselves and the aspect of the organization that is the focus for inquiry and reflection. If the representation for the second one is tacit, the re-model, as the cycle repeats, serves primarily to strengthen the existing model. However, if the representation is explicit, the re-model will typically involve modifications and elaborations. This construction of a model provides a conceptual framework for understanding the organization and its processes for learning. The third model serves as a guide for action. In guiding action, the model must necessarily be translated into concrete strategies, structures, and processes. This may involve, for example, the design of new or different tactics, ways to organize, or methods for information generation, storage, and retrieval. This step focuses on the strategies, structures, and processes for making decisions and taking actions, as opposed to the particulars of a specific decision or action.

The 7s framework that consists of shared value, style, strategy, structure, staff, skills and systems provides a systems view of an organization. A system is defined as an assemblage of interrelated elements directed towards a common goal. The seven S's are indeed interrelated, as depicted by the numerous connecting lines shown and they are directed towards a common goal: the success of the enterprise. Granted the breadth of the 7-S framework, it is nevertheless found lacking. The missing link is a most important S: synergistic teams. A synergistic team is one in which the members learn together and manifest a level of collective intelligence greater than the sum of the intelligence of the individual members. This eighth S lies at the very heart of the learning organization. These eight elements are sufficient to describe a learning organization. In fact, they could be used to describe any type of organization, but our purpose here is to focus specifically on the characteristics of a learning organization. The learning organization has two core values: excellence: always striving for the highest standards in everything one does—commensurate with the needs of the customer and the resources available. Self-renewal: creating a framework within which continuous innovation and rebirth can occur—a framework that allows the organization to adapt to a continually changing environment while maintaining the integrity of its own identity. Standing in sharp contrast to a working group is a learning team, which is a high-performance team. The members of a true learning team have mastered the five core learning disciplines: systems thinking, personal mastery, mental models, shared vision, and team learning. As a result of mastering and applying these core learning disciplines, the members of a learning team achieve a level of intelligence greater than the sum of the intelligence of the individual members. This is synergy at its best. The one characteristic of learning teams that stands out above all the rest is the ability to engage in dialogue: honest and open communication among the parties involved. This form of communication is the polar opposite of Machiavellian communication, which is characterized by deception, cunningness, manipulation and coercion. In dialogue, everything is above-board. It is apparent that an organizational structure must be designed for both stability and flexibility. The need for clarity and orderliness is self-evident, but also self evident is the need for spontaneity and responsiveness to unpredictable problems and opportunities. Thus, the opposite of stability is not flexibility, and the opposite of flexibility is not stability. Rather, these are two complementary dimensions of an effective and efficient organization. The learning organization incorporates these two complementary dimensions by establishing dynamic networks within the vertical structure. Stability is provided by the vertical structure, and flexibility is provided by the dynamic networks.

Organizational learning encompasses different levels, such as, individual learning, team learning and organizational learning. In individual learning, each person takes responsibility for learning. In team learning, teams and work groups utilize the capability of each member for the benefit of all. Teams learn to share a common approach, supporting each other in individual learning objectives, and cooperating with other teams in the learning process. Individual learning becomes organizational learning when new knowledge is transferred across unit boundaries to others in the organization that can benefit from what has been pointed out that an organization learns through its members. People may be hired because of a specific competence and knowledge, which may be gained on the job or received in formal training. Learning is an individual phenomenon, which benefits the organization entirely through the individual. Individuals learn, if the individual doesn't use the knowledge or leaves the firm, then there is no impact. The organization has learned nothing. Organizational learning needs to be systematized into practices and processes. Organizational learning can be viewed as a cognitive process or as a result. When Organizational learning is treated as a process; more attention is given to its dynamics, than whether learning results in positively valued outcomes. A three-stage model of a learning process which includes knowledge acquisition, knowledge sharing, and knowledge utilization has been proposed. Knowledge acquisition is the development or creation of skills, insights, relationships. Knowledge sharing is the dissemination of what has been learned. Knowledge utilization is the integration of learning so it is broadly available and can be generalized to new situations. Knowledge and skill development takes place not only in the acquisition stage, but also in the sharing and utilization stages. Organizational learning as a result, emphasizes performance improvement. Organizational learning is directed towards creating useful knowledge for the organization.

The capitalization of competencies in the context of organizations, and more precisely the capitalization of resources related to these competencies. We focus particularly on the way the members of an organization can use this capitalization to get new knowledge and competencies. Integrating semantic web and enhanced learning approaches in a knowledge management context promotes organizational learning. To that end, we developed an environment based on the concept of learning organizational memory. An organizational memory seems indispensable for organizational learning. An integrated organizational memory provides mechanism for compatible knowledge representation, as well as a common interface for sharing knowledge, resources and competencies. Individual learning may be supported by making organizational memory readily accessible to an organization's members. Such a memory enhance conceptual learning, it includes organization-wide communication support, access to decision-support modeling, and computerized aids for identifying and capturing individual learning experiences.

The tendency for corporate restructuring to follow environmental changes often with a long lag and occur within a fairly short period of intense, system-wide internal change suggests the presence, first, of barriers to organizational change and, second, of triggers to change. The presence of barriers to change requires little explanation sociology and psychology point to the stability preferences inherent, respectively, in social organisms, and individuals; while population ecology points to inertia as a fundamental characteristic of organizations. One of the greatest challenges which large, complex organizations face is achieving internal change in response to a changing external environment on a timely and smoothly managed basis. Our task is to explore learning systems as organizational innovations which make progress towards creating the continuously improving and self-evolving organization. Continuous improvement is a process that involves everyone, employees and managers alike. It is a process that involves the ongoing rearranging and redesigning of elements of the organization; it requires the continuous rethinking of the patterns that connect and relate different elements of the organization and connect them with the environment; it is a process that bundles together data collection, interpretation, research, experimentation and diffusion; and it involves the individual, the team and the total organization. As such, continuous improvement seeks to develop new cognitive frameworks, interpretive schemes and actions on ongoing bases. The learning process that seems to guide most of the learning systems encompasses the following: framework, concepts and tools are interrelated; the learning process is composed of clear goals, set of phases and activities; learning is a never ending process; multiple levels of learning individual, team, inter-unit, and organizational take place simultaneously; learning is anchored in solving business related problems both at the micro and macro levels; the organization provides the tools, time and space for learning to occur; and an integral part of the learning process is continuous evaluation and feedback. Organizations in the corporate world are concerned with learning if it helps them to perform better.

Therefore learning which is valuable to organizations is embodied in competencies to do things better or do different things. When an individual, group or organization has learned something it develops a competence capacity to use continuously that learning to achieve purposes outcomes. These purposes relate to the organization's current performance and its ability to learn to adapt and change for future performance. These purposes are achieved not by learning itself but by action. The learning underpins and provides quality, consistency to the action. The actions are the manifestation of the competence which the learning has created and made possible. Action has to take place over time and involve improving current performance and making effective change. Competence underpins the quality of action and enhances the prospect of superior performance. As a starting point for our discussion we therefore define a learning organization as one which develops and maintains competencies both to perform and to change the organization to maintain or improve performance. We use the term competence as a combination of knowledge, technical skills and performance management skills. The knowledge of the activity to be performed includes not only the knowledge of the task itself but a conception and understanding of how the task is connected to desired performance outcomes for example, many technically trained personnel see their job in terms of the technical application of their skills to the production process but do not conceive the work in terms of satisfying the ultimate customer. In addition, while each organization's activities are focused on a specific line of endeavor, it needs to be able to produce consistently high performance under a variety of pressures and conditions. Organizational culture and other institutions that shape the organization form the bridge between individual learning and growth and OL and growth, they affect the aggregation problem previously introduced.

Product development projects are a part of organizational learning. It does not explicitly include learning as a critical success factor. The model is developed on basis of three research streams: rational plan, communication web and disciplined problem solving. Internal and external communication, planning and overlapping versus iteration and testing and frequent milestones, do illustrate the importance of learning in product innovation. In the empirical research below, the description of project management and transfer mechanisms as well as R&D and innovation management which is related to project performance and information process. The product development process in organizational learning is of 4 types: planning activities, planning responsibilities, planning networks and sharing assumptions. Planning activities outline the project process, prepare project activities and milestones, expose tactical knowledge of the project and understand the way to achieve these objectives. Planning responsibilities outline the content of the process, in terms of individual and group responsibilities, contribute to project and understand the responsibility matrix. Planning networks understand development of the product at an early stage, and look at what is needed extra skills or disciplinary knowledge and ensure that communication is taking place at a cross-functional level. Sharing assumptions look at the pre-project planning process that provides the opportunity to dissolve some of the classic tension between marketing and R&D. This pre-project scheme prepares important knowledge regarding the tactics of the project. It develops explicit knowledge and develops team dynamics.

Systems-structural and interpretive models offer different descriptions of the cognition and communication processes which lead to organizational learning. As a result, organizational design implications derived from each view tend to stand in stark contrast to each other. However, these design implications focus almost exclusively on the ways in which organizations process information. Action typically plays a secondary role in models of organizational learning processes. Although several theorists have presented compelling arguments generally relating action to organizational learning, these works have not specified the different kinds of action that support learning from either a systems-structural or an interpretive view. In particular, researchers have not explained how different types of actions produce knowledge under different conditions of information analyzability. Information processing and communication processes lead to the legitimization of certain organizational actions and, subsequently, how patterns of action reinforce patterns of thought. By describing the integral relationship between thought, communication and action, it can improve our understanding of the link between organizational learning and organizational action. We should not think that there is a sequential logic to the components of the model. Current organizational practices support current interpretations in a mutually reinforcing cycle. Organizations that embrace ambiguity and, therefore, distrust the long-term efficacy of current interpretations, are likely to develop and maintain diverse schema. One way in which to bring about organizational change is to institutionalize the learning process and to empower people to take responsibility for their personal development. The process of institutionalizing organizational learning is complex; however, an international project group approach can be used to facilitate the transformation process, especially when it is necessary to devise new forms of market entry. What is evident is that senior managers need to think in terms of motivating junior managers to want to achieve something higher than their immediate expectations. One way that this can be done is to encourage staff to undertake continuing professional development and to adopt an entrepreneurial approach to decision-making. An international project group can be assembled in order to link the functionality of a product with its design, the application of automated production in areas of skilled labor shortage, and the negotiation of franchise operations in order to facilitate market penetration and coverage. Building trust-based relationships is time consuming and requires constant work, but is essential with respect to partnership development. In order for the transformation process to be effective, senior managers need to understand how the concept of trust is linked with control and learning.

By transformational leadership, they point to such roles as articulating vision, fostering group goals, supporting individuals, reflective critical thinking, offering personal models, having high expectations, providing shared norms and promoting collective decision making as those having a strong impact on organizational learning. Evidently, when organizational leadership openly encourages individuals and groups to search for fresh direction and to empower them to take part in critical decisions, the reservation about change is more likely to disappear. Likewise, a set of characteristics related to school culture, such as mutual support, respect, risk taking, honest candid feedback, celebration of group success and sharing needs of achievement have also been identified as crucial elements promoting collective learning. That these characteristics should nurture learning is not surprising as each tends to provide a secure internal environment for the experimentation of new ideas, new approaches and the trying out of new efforts without too much risk or consequences. Organizational learning leads to product design management that consists of situation of design in the structure of the organization and analysis and knowledge of the firm. Organizational learning helps in understanding knowledge of companies and its priorities, understand the capacity of innovation change and put maximum attention to the environment that ultimately leads to the performance of the business. Thus organizational learning is a wide area of study and involves logistics, strategy, competence, information, resources and any other sub component of organization learning & development.

CHAPTER 15

Transformation: a measure of gaining reengineering concept
(BY BAISHAM CHATTERJEE)

Transformation is an integral part of organization development where cutting costs, accessing new skills and capabilities and achieving greater financial flexibility is the key advantage. When people think of transformation, what they think of first is transformation outsourcing which is a step of deriving breakthrough growth. Outsourcing is the key advantage that small firms have to gain access to mature capabilities. Mature capabilities is an important form while another form that is used frequently is known as dynamic capabilities that are processes that help organizations remain flexible and responsive to changing environments. Organizations that have a chosen path for development and growth and gaining competitive advantage can influence the environment in which they compete by improving best practice or creating new competitive advantage.

Transformation can be classified in various types and forms: mainly into three different segments: reengineering or redefining performance; strategic transformation i.e. redefining business objective through careful description of previous objectives and in depth description and analysis of present and future platforms that would help create new competencies and corporate self renewal for operational improvement. These three segments should be utilized at any point of time during the change of behavior of an organization. All these three ideas can help solve project management problems and help to understand and meet customer needs.

Transformation also depends on the ability to support product capability and manage and support processes like facilities, HR, finance. Continuous innovation needs a high capability of technological leadership where reaching superior quality and operational excellence are the key decisions to control and to research upon. They can be brought out only in the most advanced countries that have a record of operational excellence. In the initial stage before thinking of all this, it can be found that organizations are collections of specifics like products, functions, processes and finances.

Supply chain management is also a competitive advantage strategy which is a core business process of major importance for the realization of business strategy, it can be considered to be a very suitable transformation process. As an adoption to the chapter on SCM it can be said that planning, implementation and proper consideration of customers and employees reflect qualities for flexibility of the program and quick response. Logistics is a process that is carrying forward and transforming the process of SCM including new concepts in service logistics. There are many management tools that help in understanding the modular structure and have various phases of project utilization. Marketing integration is another form of understanding and deriving organization culture. The impact of culture on strategy, understanding information and communication process help in developing business processes. Other ideas on organizational development are through operational and new SCM strategy where reengineering can help to better strategy and improve on the project selection and team development skills.

Moreover, if you look at the overall research perspective you would find that corporate strategy is a part of organization development which completely depends on market strategy and another form of growth is through qualifying and order winning criteria which is related to the manufacturing strategy's success in positive motivation of the marketing strategy. This can be brought to success in two important parts: process choice and infrastructure, which both are highly focused parts of internal and external development. There are

2 levels in a business system level and resource level where many issues were fully controlled. There should be flexibility in these ideas so that input flexibility can be understood. The links of the value chain are all dependants therefore the analysis becomes iterative. Another important area to look upon is the gap methodology where any kind of value, performance or innovation gap can be determined and action is taken to reduce it.

Globalization is an important phenomenon of organization development to establish strong brands, integrate the different functions of a business, and cut costs to mobilize processes to increase the speed of promoting different products all as a result of better CRM. The internet or the intranet has transformed an organization. E-learning is the basis by which organizations has been transformed by combining technology, services and content based on continuous assessment and collaboration. E-learning like blogs is the most significant innovation tool through e-commerce that expands their businesses globally, cut costs, streamline processes give more importance to products and customers. E-learning supports companies that are restructuring their organizations in order to gain more competence gather the strength and vision of an industry and build the internal innovation and internal mobilization ability. This helps in understanding the goals and focus on the goals through the skill gap that is the most prominent part of HR. E-learning has become a significant part of transformation as because it has cut training costs and has become a significant area of CRM growth and utilization. E-Trade, Unisys have created extraordinary e-learning projects that can not only change the phenomenon of learning through the diffusion of innovations but also look towards the world we stay. Through utilization of blogs, messages and through this diffusion of innovation create a better world for us to live in.

Considering the fact of organizational change the important processes that govern the social system are through generating, retaining and leveraging individual and collective learning to improve performance of the organizational system in ways important to all stakeholders and monitoring and improving performance. Focus is a key objective that a sharp mind always deals with that are prepared in the mental maps which can be created only through a well-developed knowledge of the business and an extraordinary understanding of critical players, relationships, events and timings affecting the future.

Small firms generally have a very dominant focus which narrows as the firm grows larger but with their potentiality for investments in their logistics but with the need to diversification and exceeding need for management they need to search for competitive advantage which small firms find easy to get. Through this focus and will another idea that comes into play is capability that in modern terms can be identified as core capability that dominates reengineering movement and managerial capability that is a symbol of decision making teams. Creative planning technique is a very important form of increasing capability where scenarios, visualization and interactive planning can stimulate strategic thinking and faster learning. Creative planning helps in developing the insight through a much better organized supply chain and creating an environment that always looks at sustainability and look at the benefit of the end customer.

Similarly a very thoughtful image can be created in the marketplace through below capacity and potential market growth that helps in developing core competences through a balance sheet or profit and loss statement on a year on year basis that is the best suggestion to understand the market share. Thus organizational development is bounded by many factors that can reduce the organization flexibility or focus and the reason for this is a reengineering project that deters growth. It is after building this organizational flexibility that a strategic plan is constructed along with a SWOT analysis and stakeholders interest.

Transformation is very necessary to understand the barriers to the process of imitation, which is a general view in emerging markets or otherwise too much of imitation retards transformation and takes away the internal strength of an organization. If the resources like ability to form quick interrelationships to improve operations and a very sophisticated economies of scale is made to research upon, then as technology improves and new technology is implemented imitation, can be reduced. Moreover SAP and inter-organizational networks implementation can help the enterprise resources so much that it automatically increases performance. Considering the relationships and importance of economies of scale many resources are a direct source of obtaining competitive advantage. Resources like oil and gas very often creates feedstock cost advantage. Apart from considering the mobilization of resources another factor that should be considered in the production cycle is producing a unique product range that creates performance advantage which can last longer is this product range is created by calculating the market gap, imitation rate and the potentiality for the business to innovate and continuously change its products. Repeated collaborations hinder growth, but ability to develop the testing equipments that would develop the ability of a product would be a great idea. Looking forward towards the change process the organizational and project level skills looks forward towards visualizing technology before looking towards TQM and process re-engineering which depends on organizational change, environmental change and technological change. All three factors are interrelated depending on the core competence of a business and how a business uses the ideas developed by their most competitive employees. It is said in earlier texts that these criteria of core competence and understanding the changes is a criteria of organizational learning if they are to increase their future performance for reconstructing and developing for organizational survival.

Internet services help in transforming the product content and are completely related to the life cycle of products. On-line books are an example of a product that makes it easier for people to know about the book by letting people know a part of it. Products can be assumed into various types like pure information product, pure physical product and pure service product which all are directed by information and product maintenance. In other words and after completing the basic factors, more research has to be put on organizational performance that consists of 5 parts: price, advertising, market share, cost and profitability which all depend on product characteristics and are directly related to product features. Country of origin has long created an important dimension for organization performance but support for imitation through this growing demand has created a huge market space that has indeed helped emerging markets mainly in the form of ICT. This momentum of selling innumerable IT packages and other manufactured products developed countries has built a strong sense of innovation through best practices and other emerging theories like BPRE and balanced scorecard. E-commerce is an important tool for creating product leadership and by allowing the customer to create their own product besides increasing effectively the products sold and provide convenience and improved customer satisfaction.

Understanding the demographics and calculating through the internal CRM links about the customer preference attributes helps in developing new opportunities for generating new sales. Internet technology or electronic commerce enables performing as a low cost channel for obtaining customer feedback and measures the customer segmentation according to demographics and preferences. Data capture and understanding service quality is more possible through this internet technology strategy where anyone can compare and assess different products as it is and the E-marketers job becomes easier to compare which is better and how he can bring a better product to the customer. DHL introduced a great example of service efficiency through its tracking system whereas Sears introduces various products that anyone can choose from.

There are many E-commerce strategies that lead to developing a collaboration space through developing long lasting strategies in SCM. Developing a marketing idea that would automatically generate a business value and take up marketing ideas like interactive marketing and selling through e-catalogs would automatically develop customer self-service in the long run and the automation techniques help customers in understanding and negotiating very easily. Materials requirement planning and inbound transportation helps in supplier performance whereas production planning and inventory management are the sharpest focus in the modern world that lead to inventory status, inventory picking. CRM helps customers to understand the demand of a product and understand the competitive advantage. To understand the internal resources of a business, it is important to understand the necessities and ability of a business unit and understand how processes are used. It is important to understand corporate ideas and understand process initiatives. Vision and mission and policy are used to understand the process initiatives that are more inevitable in sustainable development that demands transformational change. The substantive issue in organizational development deals with a particular purpose where the distribution and customer service are the key segments with more importance to corporate affairs, strategy and sustainable development.

Similarly as talked of the criteria of distribution that is related to modern channels, it is an idea of business transformation where channel strategy, a new channel mix and customer experience combine together to refresh technology and bring in new ideas that would bring preference to price and value perceptions. By taking the channel perspective it is important to understand the customers' perceptions about channels in the context of value, focusing on the cost drivers at the customer interface, looking forward toward the contact driver and choosing and designing channels based on frequency of use. Channel strategies de-clutter the business, reduce technology costs and offers clarity to customers. Thus optimization and innovation are areas of performance that helps in reducing the borders or clogs in channel marketing and thus helps in faster and more efficient channel sales. It is not only that marketing channels help in building organization profitability; other forms are the emotional intelligence that is the main technique for improving organizational EI. In fact, organizational performance and survivability hinges a whole lot on the reliability and maintainability of products than a company-wide approach.

The external cost continues to be the highest cost since these impacts directly on the customer and may lead to customer goodwill. An example of such cost on reliability was quite apparent when Ford Motors suffered irreparable loss of customer goodwill from the massive recall of Firestone tires. Similarly top management decisions may be related to reliability and maintainability management. Top management is the key in devoting needed resources and encouraging other functional units within the organization to support company-wide initiatives. Many businesses adopt strategies to minimize inventory, focus on increasing value to customers at minimum cost and also on rapid response to the needs of the marketplace. As long as competition is brisk, the labor cost-reducing innovations raise real income and boost the demand for other services and particularly for services of the kind with which quality improves through utilization of more equipment. To maintain international standards large enterprises need to specialize in order to reduce the cost associated with the increasing complexity of their operation. Businesses in order to remain more creative and outshine the risk of destroying competency by diversification, or from engaging prematurely in activities outside the technological and market paradigms with which they are familiar. In many books people have talked a lot about process innovation. As competition shifts from markets to process-innovation and to innovation in the organization of production, the most efficient process innovator makes the highest profits. As process innovation usually involves high R&D expenditure and costly new equipment, the new structure replaces familiar market competition by a scramble of investment funds.

After thinking of the idea around reliability and maintainability many people have used a decision making tool, which provides a systematic approach to assessing performance at the strategic, tactical and operational levels. Supply chain collaboration that provides information sharing, decision synchronization and incentive alignment. Performance gap can be measured through various ways that measures individual parameters to understand research based approaches and uses metrics or different statistical tools to develop through collaboration about performance metrics. The transformation process consists of three parts: technology, project and business and social capital which should have sustainability of collaboration that would understand technology improvements and understand number of awards and intellectual property that would help understand long term return on investment. Similarly the social benefits like jobs, revenue and quality of life help in increasing the prospects of collaboration. Like the adoption of new technology it is important to enhance the science, technology and innovation capability.

Similarly knowledge that is needed, wanted and helps in creating perceptions is an example of internal resource which creates a sustainable R&D and thus this helps to hold a strategic position in the marketplace and thus the need for that particular type of knowledge and its development in the market helps in creating an opportunistic environment. Knowledge marketing is a push function that conveys information and transfers knowledge from the creator of a new technology to those who would acquire it. Technological marketing is an integral component of the innovation system. Performance gaps are another important part that comes through a cross-functional team and is formed representing the performance improvement departments of the organization. E-learning, knowledge management and electronic performance are other types of internal knowledge management tools that make it evident that the performance would not fall.

The process of transformation is brought about by an OD consultant who helps in the entry into a new business which may be an intentional choice. Mergers are significant changes of organizational form which are public to commercial. At certain times crisis unfreezes the organization, then the leaders decide how it should change, and change systems to ensure that the new ways are supported. Managers recognize new opportunities and create separate business units to exploit old opportunities and often the newly transformed businesses grow at a much faster pace than older businesses. Many people have a holistic view towards developing standards in this self developed model or idea crisis is the main idea which leads to plan of change taken by the advisors which ultimately leads to reorientation for change. A major concept that would create breakthrough ideas for transformation is through ideas that can confront brutal facts and define a very difficult concept that is impossible to outrun. Another way of developing competences is through acquisitions that gives or brings in all the benefits and innovation to continue strategic revolution. A very important part of defining a strategic framework is through strategic transformation, monitoring environment, matching scenarios that would ultimately lead to inventing possible futures.

Similarly in the production process customer requirement leads to customer satisfaction known as the assignment process whereas the transformation process starts from using the economies of scale which ultimately leads to the delivered product. Similarly assessment of capabilities and needs of a customer makes us understand how successful the assessment process towards the transformation of a product has been. Everything incorporated to this type of transformation is completely related to the potentiality of the human resources. Information technology can support work transformation by delivering technology solutions that provide more flexibility in terms of how and where work is performed. Hardware is becoming more portable such that it helps in enabling more and more employees to become mobile workers. This draws employees closer to customers. Communications for perfection has become easier through mobile voice and data communications, collaborative work group technologies and conferencing facilities. Work transformation is through performance review, work analysis, technology usage and business objective. Similarly after work transformation radical organization transformation is another thing that we talk of. It requires multiple changes where strategies must be rewritten, organizational cultures realigned around different values and processes re-worked and value chains redesigned. There are few ways in which this can be done, by helping the top management recognize the scope and scale of the challenges that the organization faces, this helps people think outside the box so that the innovative competencies are sustained. However looking at the customer perspective a commitment to study potential or lead users are needed. This establishes flows of insight that enables product and service development to take place with the whole potential market in the firms mind, not the firm's current market strength.

Transformation ideas are carried out in many countries and E-governance and E-commerce are becoming an emerging part of this business transformation. This gives a clear picture of the ways to re-focus on cost, efficiency and quality and re-centralization of some strategy and control. This type of refocusing creates a new channel and innovation capacity of the business.

CHAPTER 16

Processes of Organizational design and developing internal communications
(BY BAISHAM CHATTERJEE)

Today's organizations may be seen as shifting from a paradigm based on mechanical systems to one based on natural, biological systems. Theories say that the integration of both micro and macro levels of analysis and individuals and groups affect the organization and the vice versa. To thrive managers and employees need to understand the multiple levels; ex-research may show that employee diversity enhances innovation. It is important to understand how structure and context are related to interactions with diverse employees to foster innovations. Organizational directions are achieved through decisions about structural form, including whether the organization would be designed for a learning and efficiency orientation. Organizational design rightly takes into account information and control systems, environmental needs, culture and types of production technology. A lot of differentiation strategies are also required in organization design which involves a lot of costly activities such as product research and design and extensive advertising. Undertaking marketing abilities as strength of creative employees are given the time and resource to seek innovations. Employing a low cost leadership strategy is very important for organizational design. It tries to increase market share by emphasizing low cost compared to competitors. By employing this strategy the organization aggressively seeks efficient facilities, pursues cost reductions and uses tight controls to produce products more efficiently than its competitors. Developing a long range plan on basis of an analysis of prospector, defender, analyzer, and reactor is also very important. Developing organizational effectiveness and developing multiple goals is a resource based approach. Organizational communications is dependant on all this, moreover organizational efficiency is the amount of resources used to produce a unit of output, and the ability to reach greater goals through utilization of lesser resources is considered to be more innovative. The resource based approaches understand of the bargaining position, understand the organization's decision makers and help them to perceive and correctly interpret the real properties of the external environment. The idea that measures organizational goals and performance criteria is known as the competing values approach model which consists of 4 sub-parts. The open systems model consists of growth, resource and acquisitions and sub-goals like flexibility, readiness and external evaluation. The rational goal model develops ideas that lead to productivity, efficiency, profit and sub-goals like planning and goal setting. Internal process model develops the internal infrastructure through stability and develop through communication and information management. Similarly the human relations model is different and consists of cohesion, morale and training.

Organizational communication and organization structure are interrelated. Organizational structure designates formal reporting relationship, including the number of levels in the hierarchy and the span of control of management and supervisors. It develops formation of individuals into one department and one department into many departments. It ensures design of systems to ensure effective communication, coordination and integration of efforts. It is very often described such forms as vertical structure and horizontal structure. The horizontal structure is led by self-directed teams, not individuals as the basis of organizational design and performance. In this case the process owner makes a market analysis where they have responsibility for each core process in its entirety. The core processes consist of supply and logistics process, part analysis, purchasing and material flow. People on the team are given the skills, tools, motivation and authority to make decisions central to the team's performance. Team members are cross-trained to develop combined skills to complete a major organization task. In this structure teams can respond flexibly to new challenges that arise. Other form is known as the hybrid structure that combines characteristics of various approaches tailored to specific strategic needs. This structure is used in rapidly changing environments because they offer the organization great flexibility. They combine characteristics of the functional and divisional structure. When the firm has several products to manage, it typically is organized in self-contained divisions. Organizations have to keep in touch with what is going on in the environment, so that managers can respond to market changes and other developments. A survey of high-tech firms served that 97% of competitive failures resulted from lack of attention to market changes or the failure to act on vital information. To bring external information in an organization boundary personnel scan the environment. Boundary spanners in engineering and research and development scan new technological developments, innovations and raw materials. They prevent the organization from stagnating by keeping top managers informed about environmental changes.

Inter-organizational relationships are the relatively enduring resource transactions, flows, and linkages that occur among two or more organizations. Understanding this larger organizational ecosystem is one of the most exciting area of organization theory. The models and perspectives for understanding inter-organizational relationships ultimately help managers change their role from top-down management to horizontal management across organizations. The resource dependence theory states that organizations for the supply of important resources try to influence the environment to make resources available. Organizations succeed by striving for independence and autonomy. When threatened by greater dependence, organizations will asset control over external resources to minimize the dependence. Resource dependence says that organizations do not want to become vulnerable to other organizations because of negative effects on performance. It is based on two factors: the importance of the resource to the firm, and how much discretion or monopoly power those who control a resource have over its allocation and use.

One of the resource strategies is to adapt to or alter the interdependent relationships. This would mean purchasing ownership in suppliers, developing long-term contracts or joint ventures to lock in necessary resources, or building relationships in other ways. Interlocking directorships occur when boards of directors include members of the boards of supplier companies. In a resource strategy trade associations help coordinate needs, sign trade agreements, or merge with another firm to guarantee resource and material supplies. Lobbying for favorable taxation, deregulation, tariffs or subsidies helps in making resource acquisition easier. The new model that builds new orientation partnership is based on trust and ability of partners to develop equitable solutions to conflicts that inevitably arise. Companies work for equitable profit and look at high commitment. The new model is linked with face to face or close communication and electronic linkages to share key information, problem feedback and discussion. Another new orientation is involve in the partner's product design and orientation and look at business assistance beyond the contract. Organizational form is an organization's specific technology, structure, products, goals and personnel which can be selected or rejected by the environment. A new organization always tries to find a niche. It is because of this niche that an organization grows.

The feature of service technologies with a distinct influence on organizational structure and control systems is the need for technical core employees to be close to the customer. The impact of customer contact on organization structure is reflected in the use of boundary roles and structural disaggregation. Boundary roles are used extensively in manufacturing firms to handle customers and to reduce disruptions for the technical core. They are used less in service firms because a service is intangible and cannot be passed along by boundary spanners, so service customers must interact directly with technical employees, such as doctors or brokers. A service firm deals with information and intangible outputs and does not need to be large. Its greatest economies are achieved through disaggregation into small units that can be located close to customers. On the other hand manufacturing firms tend to aggregate raw materials in a single area that has raw materials and a available work force. A large manufacturing firm can take advantage of economies derived from expensive machinery and long production runs. Again in a service technology firm the skills of technical core employees need to be higher. These employees need enough knowledge and awareness to handle customer problems rather than just enough to perform a single mechanical task. It is important to create the importance of self directed teams in a service organization. In an organization design area it is important to perform a departmental design, by understanding the departments' technology. It can be performed through a number of characteristics, such as skill level of employees, formalization and pattern of communication. Definite patterns do exist in the relationship between work unit technology and structural characteristics which are associated with departmental performance. The structure can be classified on whether it is organic or mechanistic, where routine technologies bring formal rules and rigid management processes. In a non-routine technology department management is more flexible and free flowing. The work unit technology determines the different qualities like formalization, centralization, worker skill level, span of control and communication and coordination. To look at the communication perspective, it is of 3 types' pooled interdependence, sequential interdependence and reciprocal interdependence. The first type that is pooled is the lowest form of interdependence among departments. Work does not flow between units and each department is part of the organization and contributes to the good of the organization. It is possible to perform this by means of mediating technology that helps in working independently. Reciprocal type is found in organizations that provide intensive technologies which provide a variety of products and services in combination to a client.

It is said that the balance of power has shifted to the customer. With the unlimited access to the information provided by the internet, customers are much better informed and much more demanding making customer loyalty hard to build. Moreover the concept of electronically linking suppliers, customers and partners is forcing companies rethink their strategies, organization design and business processes. Planning horizons have become shorter, expectations of customers change rapidly, and new competitors almost spring up overnight and so people need information at their fingertips. Highly successful organizations today are typically those that most effectively collect store, distribute and use information. Organization success and survival depends on how a business uses the information. Information technology can increase the power and motivation of employees by giving them complete information they need to do their jobs well, enabling them to share ideas with colleagues and finding new ways of doing things. This improves the brain power of the organization. Transaction processing system, data warehousing and data mining, executive information system and decision support system are the different ways of using information to increase the capacity of an organization.

E-commerce and other advances in electronic technology are having a significant impact on organization design. New technologies enable the electronic communication of richer, more complex information and remove the barriers of time and distance that have traditionally defined organization structures. Virtual teams for example, made up of members from divisions around the world can collaborate on project via the intranet or networks. By partnering with suppliers virtual teams pull their best minds into projects or speed a new product to the market. In a dynamic network structure a growing trend for companies is to limit themselves to only a few activities that they do extremely well and let outside specialists handle the rest. With a network organization structure, a company subcontracts most of its major functions to separate companies and coordinates their activities from a small headquarters organization. The network may be viewed as a central hub surrounded by other specialists. Rather than being housed under one roof, or located within one organization, services such as accounting, design, manufacturing and distribution are outsourced to separate organizations that are connected electronically to the central hub. The network incorporates a free-market style to replace the traditional vertical hierarchy. Subcontractors move in and out of the system according to changing needs that makes the speed and ease of electronic communication keep costs low but expand activities to market visibility. A major strategic advantage for network structures is that the organization, no matter how small can be truly global, drawing on resources worldwide to achieve the best quality and price and then selling products or services worldwide just as easily through subcontractors. The network structure can allow companies to develop new products and get them to market rapidly without huge investments. The ability to arrange and rearrange resources to meet changing needs and best serve customers gives the network organization greater flexibility and rapid response. Another significant trend is developing electronic relationships with suppliers and customers and shows the differences traditional inter-organizational relationship characteristics and emerging relationship characteristics. Technologies that enable people to meet and coordinate online can facilitate communication and decision making among distributed, autonomous group of workers. It helps in telecommuting to help individual workers perform work that was once done at home or other remote locations. They need not collaborate or stay in one roof to share information or work autonomously. Using explicit knowledge helps in developing sophisticated information technology systems that helps in developing patents and licenses, work processes such as policies and procedures, specific information on customers, markets, suppliers or competitors, competitive intelligence reports and benchmark data. When an organization uses this approach the focus is on collecting and codifying knowledge and storing it in databases where it can easily be accessed and reused by anyone in the organization.

Organizational size is a contextual variable that influences that influences organizational design and functioning just as do the contextual variables like technology, environment and goals as discussed in chapter 1. Bureaucracy determines the organizational life cycle and structural characteristics at each stage. The organizational control and how organizations determine the best means of controlling an organization has to be assumed. Large firms have characteristics like economies of scale, global reach, mechanistic and vertical hierarchy and a complex stable market. Small organizations are responsive and flexible, have a regional reach, having a flat structure, be organic and simple, have a niche finding and a start up role to be played by the entrepreneurs. Countless small businesses have sprung up to fill specialized niches and serve targeted markets. The development of the internet has provided a fertile ground for the growth of small firms. Small organizations have a flat structure and a organic, free-flowing management style that encourage entrepreneurship and innovation. Moreover, the personal involvement of employees in small firms encourages motivation and commitment because employees personally identify with the company's mission. Small companies, however, can become victims of their success as they grow large, shifting to a mechanistic structure emphasizing vertical hierarchies and spawning organizational men rather than entrepreneurs.

Another approach to creating a big company/small company hybrid is called the front/back approach. Rather than dividing the company into separate businesses, each with its own products and customers, the company is divided into units with different roles. The back part of the organization focuses on creating and producing products and services, while the front focuses on integrating and delivering products and services to customers. This approach is becoming increasingly popular in financial services companies and multiple-product technology firms. Full-service global firms need a strong resource base and sufficient complexity and hierarchy to serve clients around the world. The development of new organization forms, with an emphasis on decentralizing authority and cutting out layers of the hierarchy. Combining with the increasing use of information technology making it easier for both large and small companies as well as capturing the advantages of size in areas like advertising, purchasing and raising capital. In the organizational life cycle the few stages are entrepreneurial that shows creativity and arouses the need for leadership, collectivity stage that has provision for clear direction. The need for delegation and control, formalization stage that has an addition to the internal system, and lastly the elaboration stage that needs to deal with too much red tape, needs revitalization, develops teamwork which all as a total streamlines small company thinking and shows continued maturity and decline. It is a new sense of collaboration and teamwork, social control and self discipline and learning to work with the bureaucracy that solves the red tape crisis.

Market control occurs when price competition is used to evaluate the output and productivity of an organization. The use of market control requires that outputs be sufficiently explicit for a price to be assigned and that competition exist. Without competition, the price will not reflect organizational efficiency. Even government firms are turning to market control. Companies are finding that they can apply the market control concept to internal departments such as accounting, legal practices, and data processing and information services. In a different way clan control is the use of social characteristics, such as corporate culture, shared values, commitment, tradition and beliefs to control behavior. They generally require shared values and trust among employees. They are important when ambiguity and uncertainty are high. Organizations may use a bureaucratic, market, and clan control to best meet the needs of various departments and the total organization. The most recent innovation is to try to integrate the various dimensions of control, combining internal financial measurements and statistical reports with a concern for markets and customers.

Organizations must run fast to keep up with changes taking place all around them. Organizations must modify themselves and bring change management paradigms from time to time or with time keeping a short term or long term goal in mind. Large firms must find ways to act like small, flexible organizations. Manufacturing firms need to reach out for new computer integrated manufacturing and service firms for new information technology. Today's organizations must poise themselves to innovate and change to prosper and survive. Taking the perspective of innovation and change it can be found that the powerful forces associated with advancing technology, international economic integration, the maturing of domestic markets and the shift from communism to capitalism have brought about a globalized economy that impacts every business from largest to the smallest creating more threats as well as more opportunities. To manage these threats and opportunities many firms are forming self-directed teams that enhance collaboration and communication, streamlines supply and distribution channels, and overcomes barriers to time and place through E-commerce. It is important to focus on core competencies, knowledge management and ERP that facilitates the network and helps the business keep pace with what is happening within and outside. To develop an organization, we have to manage technology changes that are in an organization's production process, including its knowledge and skill base and enable distinctive competence. Techniques for making products or services produce greater volumes of products in respect to the condition of the environment both internal and external. They include work methods, equipments and workflow. New products are normally designed to increase the market share or find new markets. The organization and the environment are interrelated with the internal or external environment consisting of suppliers, professional associations, consultants and research literature, customers, competition, legislation and labor force that should communicate their ideas to the organization from which the needs can be predicted and the internal creativity and perceived problems can be understood. The next steps after this are utilizing the adoption, implementation and resource utilization process. Organizations face a contradiction when it comes to technology change, for the conditions that promote new ideas are not generally the best for routine production. An innovative organization is characterized by flexibility, empowered employees, and the absence of rigid work rules. For adapting to a chaotic environment, an organic free flowing organization is the best choice. It is attributed to people's freedom to create and introduce new ideas and encourages a bottom-up innovation process. Middle and lower level managers bring the main idea for development and experimentation. The technical or product champion along with the management champion allocates resource and acts as a supporter and sponsor to shield and promote the idea within an organization. He sees the potential application and brings in ideas to add resources to it. New product success is a true communication phenomenon and depends on how the three processes like technical completion, commercialization and market success in carried on to bring it to a success. These three processes look at achieving technical objectives, full-scale marketing and earning economic returns. Successful innovating companies paid more importance to marketing and look more at outside advice, work-in-house and create greater authority.

Horizontal linkage model consists of specialization, boundary spanning and horizontal linkages. The key developments in new product developments are R&D, marketing and production by means of highly competent personnel. Boundary spanning is undertaken by R&D personnel and marketing personnel who share ideas and information. In a dual-core approach that consists of administrative core and technical core where innovation can occur in either core. The administrative core is above the technical core and its responsibility includes structure, control and co-ordination of the organization and concerns the environment sectors of government, financial resources, economic conditions, human resources and competitors. The technical core is concerned with the transformation of raw material into products and services and involves the environmental sectors of customers and technology. Technically innovative firms should reorganize frequently or must suddenly cut back to accommodate changes in production technology or the environment. Technically innovative firms may suddenly have to restructure, reduce the number of employees, alter pay systems, disband teams or form a new division. In the horizontal organization, managers and front-line workers need to understand and embrace the concepts of teamwork, empowerment and co-operation. Everyone throughout the organization needs to share a common vision and goals so they have a framework within which to make decisions and solve problems. Change management is brought out by commitments like initial contact, awareness, understanding, decision to implement, installation and institutionalization and all this is supported by commitment, acceptance and preparation.

The Carnegie model consists of management coalitions that are needed during decision making for two reasons. Organizational goals are ambiguous and operative goals are inconsistent that makes managers disagree about problem priorities. Managers do not have the time, resources or mental capacity to identify all dimensions and to process all information relevant to a decision. The Carnegie model says that whether it may be problematic search or satisfying, they say that it produces a search behavior that managers typically adopt. The Carnegie model applies when there is dissension about organizational problems. When groups within the organization disagree or when the organization is in conflict with constituencies like government regulators, suppliers, unions, bargaining and negotiation are required. Once bargaining and negotiation are completed, the organization would have support for one direction.

Communication in organizations should provide accurate information with the appropriate emotional overtones to all members who need the communication content. Neither too much nor too little information is in the system nor is that it is clear from the outset who can utilize what available. The communication process is paradoxical and contradictory and permeates organizational life. Any text book gives a clear picture of the different communication processes. Making a recent analysis, through research data the most prevalent major changes in respondents communications functions in the past three years have been the establishment or restructuring of an in-house communications team, and or an increased emphasis, recognition and acceptance of the need for, and or focus on managed communication activity, particularly for internal communication. Where this is in evidence, more senior, better trained people have been recruited to take on the associated responsibilities. While teams may, in many cases, be smaller, they are producing better work. More active communication and greater use of outside service specialists have also arisen due to increasing awareness of the need to communicate. Channels have been reviewed due to a perceived need to be more discriminating in the targeting of information. Communication activities are becoming more focused on particular stakeholder groups, and the emphasis chosen is specific to particular industries and sectors. This has often been in response to pressures from external groups, such the media, regulators, and pressure groups. Global electronic communications will change our lives more slowly than many commentators have predicted. However computer Internet working is disrupting the comfortable power balance away from corporations and organizations towards individuals. Communication theory is centered on process rather than information, but the changes in communications technology are affecting both the process of corporate communications and the balance of power that has resided with those who hold information. This power shift means that organizations can no longer manipulate their information assets that are their useful and valuable information resources—to ensure a largely one-way communication, and pay mere lip service to the notion of corporate communications. Analyzing how world class companies became world class, suggests that one of the major dynamics of a successful organization, in performance terms, is through communication excellence. He advocates that a way of cultivating trust is through open communications, where comprehensive information helps to improve problem solving. In this way, motivation and identification are high in the profile of a transparent and accessible organization. In his opinion, everything imaginable should be communicated as it is important to supply regular information about the corporate strategic excellence position, thus promoting and giving substance to the corporate identity. Corporate vision and communication objectives are at the apex, underpinned by market and financial indicators which are considered essential elements for success. This success is based on the operational determinants of customer satisfaction, flexibility and productivity which, themselves, are founded on high standards of quality and delivery, reduction of cycle time and minimization of waste. The pyramid concept, though seemingly a rather rigid construction, is a useful model to describe how objectives are communicated down to the troops and how measures reach the various levels in the organization. However, the consequent feedback must be flexible enough to capture the changing organizational dynamics.

Organizations should also have local consideration in the change factors so that a change in one division does not have organization-wide impacts. Moreover sometimes organizations are over determined which means that there are multiple areas to ensure stability. Personnel selection, training and reward system are designed to lead to stability. There is individual and group inertia and any change should be determined and backed with forecasted data, because sudden changes can threaten occupational groups within an organization. It can threaten the established power system and sudden change scatter the ideas behind present allocation of rewards and resources. Organizations enter niches in an environment. The niches contain the resources for the organization and are likely to contain other organizations fighting for the same resources. These organizations that survive are the only ones that coexist with their competitors. Innovations within an organization are not random: innovations occur in relation to the past and present conditions of the organization. Innovation is of three forms programmed innovation that is planned through product and service research and development, non-programmed innovation occur when there is slack in the organization in the form of more resources available than are presently needed. These are called non-programmed because the organization themselves doesn't know when such extra resources will be available. To determine the adaptation of a new technology depends on such factors like whether the technology has sound scientific status, the product has less risk and uncertainty, whether the product originates within the organization, then the product would get more credibility. The timing of the innovation and the technological policy can also be understood and that definitely reflects the way an organization is managed, how well communications take place. Linkages have brought about inter-organizational networking. Organization theorists began paying attention to IORs when they started looking beyond organizational boundaries into organizational environments and when urban sociologists began to recognize urban communities as networks to organizations. This type of networking consists of linking all organizations by a special type of relation and is constructed by finding the ties between all organizations in a population Pair-wise or dyadic inter-organizational network, set and a network.

Networks are the total patterns of interrelationships among a cluster of organizations that are meshed together in a social system to attain collective or self-interest goals or to resolve specific problems in a target population. A new approach or Aldrich approach is different and it does not concern with groups or target goals but focus instead on linkages, such as financial or other resource transactions. Personnel or client flows should be the linkage in the Aldrich scheme. An empirical analysis would probably find a close overlap with the other approach that stresses the network in areas such as social or health care delivery within a community and would include all of the organizations in that network of service delivery. Since organizations linked by resources flows in the health care area would ultimately also be identified following this framework. The complexity of the relationships is indicated by the fact that frequent interactions do not necessarily mean highly formalized or co-operative relations. It has been found that co-operation and conflict can exist in the same relationship. The analysis of the IORs is complex. Not only does an organization like the main organization through which sub-units are formed, but have multiple other sets of relationships. Each organization must purchase goods and services. Many departments are also involved in finance assistance program with linkages to federal, state, and local organizations as well as citizens groups. This form entails the creation of a new organizational entity by organizations joining in a partnership. The joint venture is a means by which illegal mergers can be avoided but yet permits joint capital investments on the part of the organizations involved. Among profit seeking organizations, oil and gas exploration efforts are a common form of joint venture as the participating organizations which both competitive and symbiotic relationships among themselves, seek to reduce environmental uncertainty and reduce the risk for each participant. When the analysis is focused on sets or networks, even more complexity, is introduced since the focus can be on the set or network, as a focal organization within the constellation and other organizations. A study examined the linkages with environmental elements that modify power relationships within an organization set. It is more difficult to establish or maintain relationship across distances for both organizations and individuals. It can also be noted that the type of unit involved in an IOR interacts with the spatial issue. Modern communication techniques permit rapid information flows across space, but clients or staff members would be more difficult to transfer. An increase in the number of organizations in a relationship affects dependencies, domains, and the potential rewards or resources for participating in the relationship. Many ties reduce the likelihood for each of the ties being strong, so that a greater proportion of linkages in a large network would be more superficial than in a smaller network. It would appear that a large set of network would have the potential for dissipating resources and actions, but it could also lead to a situation in which there were many alternatives for an organization, in terms of resource acquisitions, client flows and the like.

There are 3 basic reasons for IORs. The first reason is the procurement and allocation of resources such as facilities, materials, products and revenues. These are crucial for organizational survival. The second reason is to form coalitions for political advocacy and advantage, while the third is to achieve legitimacy or public approval. The last reason can be seen in instances of interlocking boards of directors among cultural organizations and corporations. In the exchange bases the exchange idea incorporates the notion that organizations must acquire resources and the exchange is the major mechanism by which this occurs. It is a form of bargaining to acquire resources in another organization. The exchange basis emphasizes the importance of resource acquisition for the organizations involved. It also shows rationality, as the organization tends to maximize their goals in interaction. All the problems involved are considered in exchange interactions and spheres of organizational actions. A study of business firms found that they tend to continue relationships with their auditing firms over long periods of time. Resource exchange has far reaching consequences. Perception of resource dependence is an important spur for IORs. Resource dependence is a powerful direct determinant of IOR communications, resource transactions and consensus. Monetary transactions and client referrals involve different patterns of communication.

The flow of resources varies in their intensity, so that the quantitative idea of interaction can be understood. In the case of resource interdependence a situation in which two or more organizations are dependant upon one another for the resources each has access to or controls the basis of resource interdependency. Resources take a variety of forms, such as inflows and outflows of information, money and social support. Organizations that have intersecting domains tend to be more interdependent. Equipment, meeting rooms, funds, facilities and personnel can be considered as other resources. Similarly the level of resource investment required of the organizations determines the IORs and intensity of relationship. The more intense the relationship, the more important it is for the organizations involved. Intensity varies from casual to consuming. The former makes little difference but the latter has the ability to consume the organization if all its efforts involve inter-organizational relationship and the relationship uses up all the resources. Intensity is sometimes confused and is a bad mix, since frequent interactions can be casual, but a really intense communication can be infrequent. Dyadic relationships, networks and organizational sets can all vary in the degree of intensity. They become more intense during crisis situations opposite in case of non-crisis situations. The impact of intensity is both network-wide and organization specific. In an organization, records and policies can provide the framework and content for decisions to be made. This provides continuity and looks at so that past ideas are retained. Policies and advancement based on performance are to be followed. In addition role specialization and standardization, with related job description shows that policies are followed. If people are advanced up through the system, their experiences would be quite common and they will react in similar ways. Bureaucratization also helps the retention process and looks at advancement based on performance, which ultimately leads to continuity and is probably the most efficient form of administration and all organizations will move towards this form if they seek efficiency.

In a different perspective taking the case of selection within organizations and work groups, internal diffusion, imitation, promotion and incentive system may be selective in ways that enhance fitness or are simply irrelevant. Managers can often introduce positive internal selectors, by favoring the strategic direction and then the elements of organization design like logic, goals, scope and competitive advantage. Management and business strategists usually focus on selection systems that improve fitness, whereas an evolutionary approach alerts us to the possibility that many selection systems are irrelevant or not tightly connected to environmental fitness. They preserve organizational diversity that is not tied to current environmental conditions. Three types of internal selectors contribute to the loose coupling of internal selection and environmental fitness: which pressures towards stability and homogeneity, the persistence of past selection criteria that are no longer relevant in a new environment and the willingness of some organizational founders and leaders to accept a low performance threshold, Pressures in work groups and organizations often encourage internal stability and cohesion. Frequent interaction between members leads to positive reinforcement of interpersonal behavior that is rewarding for the people involved and to the elimination of incompatible behavior. Such shifts in choices or attitudes within a group have been explained by social comparison, self-categorization and network influence theories. The self-reinforcing process contributes to organizational stability, but also leads to competency traps that inhibit the discovery of potentially adaptive alternatives. Members may simply continue doing what they know best, rather than searching for more effective options. Investments in human capital specific to a particular organization, psychic income from association with the organizations, and the costs of switching to another activity make some founders and leaders less sensitive to low organizational performance than others. Accounting and information management system create categories that channel and document certain activities, directing members attention towards them and away from undocumented activities. Organizational memory also inheres in physical resources such as buildings and machines. Organizations often cling to traditional ways or display a reluctance to trust outside information. Decisions on which variations to copy are clouded by ambiguity in outcomes observed from a distance. Tacit knowledge embedded in an organization's routines may mislead outsiders into imitating the wrong variations. Hostility, mistakes, incompetence and a willingness to learn also impede diffusion. New organizations require that investigators identify when social entities begin. As goal directed, boundary maintaining, activity systems, organizations become new social entities that have a taken for granted presence in a society. The boundary between pre-organization and organization identifies the criteria for understanding existence: intentionality derived as stated goals, mobilization of necessary resources, coalescence of boundaries, such as through formal registration and naming of the entity, and the exchange of resources with outsiders. The new organizational knowledge they carry may thus transform an existing population to create a new one. They have created competence enhancing or competence destroying innovations. The first criteria is based on adaptive innovations or involving substantial improvements that build on existing routines. Organizational boundaries can be transformed in 2 ways by expansion or by contraction. Through mergers and acquisitions organizations might expand its boundaries in order to take in other organizations or contract boundaries through divestures. Activity systems in organizations are the means by which members accomplish work, which can include processing raw material, information or people. It consists of bounded sets of interdependent role behaviors that are contingent upon the techniques used.

The lack of convergence on a dominant design—an agreed upon architecture and set of components constituting a product or service—in new populations constrains the perceived reliability of founding firms by increasing confusion about what standards should be followed. Not only must founders convince skeptics, of their organization's staying power, but they must also fend off organizations offering slightly different versions of their product or service, creating confusion in the mind of constituents. During radical innovations a struggle occurs between contending designs, this ferment ends when a dominant design emerges for the core sub-system. Convergence towards a accepted design is facilitated if new ventures find it easy to imitate pioneers rather than seek further innovation. New technologies display demand side increasing returns. If the adoption of a technology permit adopters to gain experience and perhaps improve it, then seemingly then insignificant events may give one form of the new technology an initial advantage that competing technologies cannot overcome. The fortunate technology, becomes better than competitors and improves more rapidly and thus appeals to a wider set of potential adopters, enjoying still more chances to improve. The process is path dependant which can be illustrated by well documented historical cases, showing how an early established technology gained such an overwhelming advantage that subsequent superior potential technologies were locked out: the design of typewriter keyboards and conflict over alternating versus direct current. Implicit agreement on a dominant design, common standards and the inter-firm movement of personnel increase the level of shared competence. Early automobile manufacturers almost always began with under capitalized firms, as established financial institutions would not risk their assets on an unproved product with no clear market. The selection and retention of a dominant design may simply shift the competition to alternative technological trajectories within the design, rather than ending the period of ferment and experimentation surrounding competition between competing designs. Technological innovation constitutes a major catalyst for the creation of new organizational forms. Rarely do single key events generate new organizational populations, based on a technological breakthrough. Instead from an evolutionary view, technological innovation typically results from a cumulative series of interrelated acts of variations, selection and retention, eventually culminating in commercial applications. Because a population's product or service often comprises part of a larger symbiotic system of components, its evolutionary path depends on changes in other populations. Many innovations are related to some aspect of a technological system, which are thought to consist of core peripheral subsystems. Most micro-electronic devices are sold as components of more sub-systems., unlike biotech products that are sold directly to consumers. Incremental change is relatively stable with respect to core sub-systems but it may be quite dynamic with respect to innovations in peripheral sub-systems. Individuals and organizations can cause temporary uncertainty by creating peripheral subsystems and complement the core technology. Technological breakthroughs might go beyond current organizational knowledge but still allow existing populations to participate in a new community. Competence extending innovations look at new opportunities and look at complementary ventures. They are not a straightforward extension of the current routines and competencies. At the same time however these opportunities are not direct threats to their existing business pursuit and competencies. Organizations in the modern world are more complex and are searching for new markets to gather new competencies. From the 1990s organizational development has its major developments through the complexities, different models developed, the learning process, pursuit of self exploitative goals and exploit future competencies. Thus organization developed the needs of more theory analysis, psychology understanding and future experiments.

CONCLUSION

There is success as well as debacle in all organizations. The chapters that I described improve the processes and resources, whereas the conclusion would describe organizational decline, advantage of modern communications that a business can provide to society and how all the ideas that has been described in the chapters, fit in together.

Organizational decline occurs due to a lack of good information, misallocation of resources, environmental uncertainty, lack of correct action and lack of effective reorganization. Decline of a business occurs due to the threats that a business can't tackle like deficiency in the performance of marketing channels, problem in allocating right amount of raw material at the right time to the organization and excessive market entry threats and extensive globalization that has created competitive threats. To manage this best, a business should try to solve through the theory implied through the Porter's five forces, uncertainty analysis as in the book 'competitive advantage' by Michael. E. Porter and understanding the industry segmentation. But the cycle that creates and leads to this decline is: deficient potential slack and performance, further deterioration of potential slack and performance, strategic extremism and vacillation, neutral or buoyant environment, satisfactory working capital, marginal potential slack and performance and finally in the death struggle which is lead by strategic extremism and vacillation, sudden environmental decline, sharp deterioration of slack and performance and failure. Failure of a business is also dependant on changing market needs and whether an organization can change its design and effectiveness with the market needs. Performance in a cluster stops organizational decline but increases the competition within the cluster although associations formed within the cluster, guides any business within the cluster. But clusters are not formed all the time and very often the organization exists as an individual identity in that particular country and needs better communication skills like a GIS or satellite system that would drive production faster and thus communicate without any gap in communication and information sending from any department or any unit of the organization. Using different benchmarking tools to manage the different departmental levels and hierarchy levels definitely stops any outside economic policy, rules, or market changes helps an organization to grow. Finance is not the core of a business and the core subject of a business is only managing the stock markets and the other ideas that keeps a business afloat are the ideas that I described in above. Businesses are getting more complex and to drive forward the idea of radical innovation, any business should be cautious about the history of its products markets, needs and changes of likings of customers from time to time, research on how markets grow, how economies flourish and how changes can be described to bring a fit in all these ideas and conclude that the business would not decline but keep on performing for a long time. Modern communications are through the internet and satellite systems as well as looking at the best technology news and developments that would not let any production gap create major threats. An R&D unit and understanding any recent technology development and bring in internal developments by means of consultants, is the best way of communication in the modern world. An efficient telecommunications by using 3G, wireless, PDAs and other Windows vista tools definitely make things clearer and through visual interaction it doesn't leave the time gap and hence utilizes money, time, resources, efficiency and does not leave any production gap.

All the different descriptions in all the chapters would help form a very creative book with many broad focused ideas where much research hasn't been made yet. As in the first chapter it is said that many short term goals together create very prominent long term goals. But as economies are booming with thousands of both similar and different businesses in every country, the organization that can forecast future and has talent is sure to win. This futurist outlook needs a visionary leader to compile the goals created in various departments by taking advantage of the organizations past history, and a 2-3 year goal determination and workload setting can create a clear concept of core competence and determines through the usage of technology tools and organizational learning that the business would grow in the next 40 years, keeping in mind and calculating the problems the economy may face due to any economic downturn. Transformation is the change activity that can suggest how the business should change with uncertain times and competition. Communication system helps in making an analysis of the advantages created in other nations and determines how a similar business in another country can improve its transport, warehousing and logistics and how to improve it using GIS and other very recent communication tools. This is a part of supply chains and to do all this and focus more on quick profits through easier utilization of resources and quicker transport, businesses are shifting to E-commerce sites for quicker order rates and stand out to be ahead in the race. Small businesses and retailers are generally using E-commerce, but even manufacturing firms can shift to an E-commerce web site or sell their products to large retailers and advertise them through internet ads that can save costs, time and increase talent and efficiency to set future goals. To create these goals there needs to be a source of motivation that can help build talent through continuous appraisals and incentives, by means of which the business would never be left behind. Customers are always the target and interest, path goals, stretch goals and customer relationship management are developed in this perspective so that any type of goal is attached to the type and demands of the customer. Product prototype strategy and investment strategy both need high competence goals like how to produce prototypes that would keep on serving and the rate and research required. The components of the product and how to make its efficiency better. All this is a result of extensive research and a concrete time to time goal setting based on similar products in other organizations and achievements of the product in the particular organization. Investment strategy can be in manufacturing oriented organizations which is very long term, like completing a 50 year term project which is also very long term, as well as investing in SMEs which needs a short term objective but many multiple objectives. Project completion needs time goals and transport goals as well as resource goals which needs high expertise. Large projects need the presence of latest technology tools in ERP to utilize and create further resources and manage the project and the skills and money in the best possible way. Vision is a result of goals that is, it is the next focus to be successful and needs high environmental dynamism and ability to understand both the internal and external environment without any lag and thus helps explore the quality domain and explore other strategy and technology tools to explore new markets and lead the organization. This type of skill is often required, after a period of uncertainty, that has created a technology and efficiency gap. This helps create a visionary company that can keep on succeeding by creating a relationship and linkage between vision, goals and innovation. Many kinds of visionary companies can be created with different visionary models and study of history which is mainly practiced in organization behavior. Mid sized businesses have the highest potentiality to use this linkage with the result of a visionary organization. Incentives help manage the internal system to bring performance and system of ISSON and incentives through e-governance and e-democracy can regulate any patent, new ideas, motivation mechanisms, leadership in any mid sized business consisting of around 800 employees. A lot of research has been done on the new incentive programs and the 3 incentive programs that would create a strong internal as well as external focus to measure and control the work transformation and the group involvement in an organization are performance based incentive, social incentive and investment incentive. Bringing other benchmarking, best practice and balanced scorecard tools like BPRE, JIT, TQM and many other processes determine the flexibility and helps build different types of benchmarking activities like co-operative benchmarking and competitive benchmarking, which together calculates the production control measures and all the management tools that would help build success. Balanced scorecard tools help in efficient calculation of how to improve on the previous results, and ideas like strategy maps have made things easier for a business. It has been found that new technology and new product development is an integral part of R&D and in depth future product study and radical innovation. Searching new capabilities is an idea of gathering the competence ideas which are a source of forming virtual businesses and e-marketplaces. E-commerce study and development is the ideal tool that is making things easier. It is very important to have an internal IT infrastructure for fastest communications internally which otherwise would be impossible. Enterprise resource planning helps in generating real time data, enabling linkages across functional area and helps in developing processes and understanding environmental information systems. Moreover CAD, CAM, ECM are other factors that help in developing and designing ideas to communicate faster in a group. It is the most advanced tool required in a high tech business. Ideas have also been based on the most recent organizations based on supply chains which is to support ERP, and the various structures formed to support that are SC integration, SC performance and testing the modularity of the supply chains. Supply chains are needed in any organization and to keep it up to date and perfect is very important. Interrelationships and economies of scale are an important study from the ideas in competitive advantage. There the many outcomes through the study of interrelationships as between network externalities and information asymmetry or between corporate identity and reputed brand where a SWOT and advantage analysis of each of the identities helps identify how to make the relationship stronger, by understanding the common points of each and find out strengths and further derived relationships. Similarly economies of scale and scope is very often linked to resources and its utilization and better helps in understanding interrelationships through a balance of resources. Linkages among value processes, channel linkage and Likert scales describe how even more than two elements form a relationship and derives ideas that would lead to the future. These ideas that are formed through the linkage is a key strategic resource that can be combined with other such results to form the net value driven outcome or result. Similarly all the points that has been described needs performance measurement to analyze the project goals to help perform better. All the technical analysis are lead by performance measures that are driven by accountability, financial, technical or control environment. To make this happen the environment as well as the business should be monitored first, which would say, yes this has happened and how history has shown that possibility to occur and the various records in this field. There have been many developments in performance measurement like the non-financial performance measure or the activity based costing which are the most recent developments of today. Similarly competitive benchmarking is related to performance benchmarking and is composed of cost, process and a strategic reference framework.

Competitive benchmarking measures the outcomes and weaknesses and strengths of the internal and external environment in order to understand the business as a whole and understand how the business should perform. Similarly any form of competitive strategy understanding should be added by competitive benchmarking to give wheels to a business to measure and compete successfully. Deming's ideas and other technology ideas are looked upon as a predominant part of competitive benchmarking and are necessary components excluding the new ideas like pre-analysis and effective benchmarking. The AHP or analytical hierarchy process is another measuring tool that compares the overall performance of different manufacturing departments through multi-attribute financing and non-financial performance criteria. It helps understand the manufacturing priorities and the competitive strategies. AHP helps in understanding any outsourcing decision problems in a criteria, sub-criteria and decision alternatives. Quality function deployment, and a multiple criteria decision making tool are other AHP factors that all depend on many criteria and functions to judge and create plans.

To look at the most modern tools: EDI, ICT diffusion, e-servicescape, and M-commerce are the recent E-commerce technologies. E-commerce helps in utilizing the market gap at a much quicker pace and thus creates a much faster knowledge base to understand and keep in record and thus use other internet and technology tools to utilize the resources and create a framework before it is implemented. ERP very often helps E-commerce in modernizing the way to utilize the resources and build a order and customer system that would help locate potential and interested customers, thus tracking the way to make profits at a much quicker pace. Moreover E-supply chains and other E-tools help perform better the E-commerce efforts of a virtual businesses, and thus virtual businesses are the most widespread and emerging business to be concerned about. After evaluating the strong positive effect of implementing E-commerce, we have to look at core competence and organizational learning. There are different types of competences, which are all parts of core competence. Finding out the future of technology or product development, environmental orientation of a business is a key success, which is thus the most important part and the most important challenge of organization development. A difference in opinion may cause hazards and make it unsuccessful to implement the above quoted factors. At first a business should understand how it should build a relationship with its core ideas and organization learning, in order to measure, use technology and vision and find out what type of structure the business should form in the future. Without understanding the concrete structure in mind all the other tools cannot be analyzed and performed in the value chain before other structures formed through the dependence on E-commerce is materialized. Organizational learning depends a lot on core competence, performance and other technology tools. Thus creating a business scenario through previous business history, helps understand how the tools can be related and put in order according to the usage of the particular business, in the particular segment, of a particular country. They vary from time to time with usage of more technical tools and the market exchange rates (which although not a major, but an important function). Talent is a key source for rating this, and a group decides on how they would be organized, in the different departments. Imitation is a key area that deteriorates the ability to gain or make profits and for this a business needs transformation. Transformation to a virtual business or using technologies that are only developed by independent and very famous research labs helps transform and modify the core of a businesses. Virtual businesses leaves the interior unknown to people for a long time, and the speed of utilization of space becomes so quick that it is impossible to imitate and the business becomes a transformed business. Examples in books, magazines, journals and websites provide learning through competitive advantage and e-learning provides the greatest source of competitive advantage. This helps in getting towards globalization with very less percentage of imitation. Virtual businesses help drive CRM and channel strategies without the bargaining power of buyers and suppliers having much effect on the sales of a virtual business. This chapter on transformation has described the theories, but the major strength of a channel is being different from others and being automated, so that there is no mistake. Communication usage and the way of handling the communication services and its different forms through research is making it easy to get access to real time data and the concerned person can transform work in a more effective way. Similarly organizational design and communications design with the help of models, where the organization would lead to or the direction of an organization. Thus ideas like pooled interdependence, boundary spanners are the modern tools that gives a new theory to organization development. I have included this as the last chapter, because organizational designing gives the structure and foundation to the other theories that help increases profits. Definitely there are many new ways of organizational designing from many books, but I have provided a brief description of what organization designing should be like, the idea boundary spanners has been described very often and ideas in inter-organizational communication have been founded. Thus this book would be very creative with the main idea behind organizational designing and finding out factors that would drive profits further in the high-tech world and developed world scenario, by improving the internal processes.

REFERENCES

BOOKS

1)Richard L. Daft(2001), Organizational theory and design, South-Western College Publishing

2)Howard E. Aldrich and Martin Ruef(2006), Organizations evolving, Sage Publications.21, 49, 252-253

3)Richard H Hall (1996), Organizations: Structures, processes and outcomes, Prentice Hall. 169-253.

4)Michael E. Porter (1985), Competitive advantage, Free Press. 48, 50, 85, 116-117, 168-169, 317-363, 383-415.

5) Kenneth C. Laudon and Carol Guercio Traver (2002), E-commerce: Business, technology, society, Pearson Education Inc.

PAPERS

1)Marco iansiti and Jonathan West(1999), Technology Integration, Harvard Business Review.6-18

2)Henry W. Chesbrough and David J. Teece(1999), When is virtual virtuous?, Harvard Business Review.35-45.

3)Gary P. Pisano and Steven C. Wheelwright(1999), The new logic of high-tech R&D, Harvard Business Review.60-79.

4)Marco Iansiti(1999), Real-World R&D, Harvard Business Review.95-106.

5)Charles R. Morris and Charles H. Ferguson(1999), How architecture wins technology wars. Harvard Business Review.121-140.

6)Walter Kuemmerle(1999), Building effective R&D capabilities abroad, Harvard Business Review.175-185.

7)Benham Tabrizi and Rick Walleigh(1999), Defining Next—Generation Products, Harvard Business review.202-219.

8)Carliss Y. Baldwin and Kim B. Clark(2000), Managing in an age of modularity, Harvard Business review.7-15.

9)Nirmalya Kumar (2000), The power of trust in manufacturer-retailer relationships, Harvard Business review.106-115.

10)Marshall L. Fisher(2000), What is the right supply chain for your product, Harvard Business Review.138-147.

11)Richard Normann and Rafael Ramirez(2000), From value chain to value constellation, Harvard Business review.192-204.

12)James P. Womack and Daniel T. Jones(2000), From lean production to lean enterprise, Harvard Business Review.222-231.

13)David Rejeski and Carly Wobig(2002), Long term goals of governments, MCB University Press Ltd. 14-16.

14)David Emsley(2003), Multiple goals and managers job related tension and performance, Journal of managerial psychology.347-349.

15)Bernd Heinrich (2005), Transforming strategic goals of CRM into process goals and activities, Business Process Management Journal. 713, 714.

16)Peter Kueng and Peter Kawalek(1997), Goal-based business process models: creation and evaluation, Business Process Management Journal. 22-26.

17)Goran Lindahl and Nina Ryd(2006), Clients goals and the construction project management process, Facilities. 150-153.

18)Roberta Mokrejs Paro and Claudio Bruzzi Boechat (2008), Collaborative governance: new roles, models and strategies for the firm. Strategic planning and millennium development goals in Brazilian companies, Corporate Governance. 535-542.

19)David C. Leonard (2007), The impact of learning goals on emotional, social, and cognitive intelligence competency development, Journal of management development. 112-120.

20)Kevin S. Groves (2006), Leader emotional expressivity, visionary leadership and organizational change, Leadership and Organizational development journal. 567-577.

21)Tony Manning and Bob Robertson (2002), The dynamic leader-leadership development beyond the visionary leader, Industrial and commercial training.139-143.

22)Robert S. D'Intino, Trish Boyles, Christopher P. Neck and John R. Hall (2008), Visionary entrepreneurial leadership in the aircraft industry, Journal of Management history. 41-49.

23)Tony Morden(1997), Leadership as vision, Management Decision. 668-672.

24)Pentti Malaska and Karin Holstius (1999), Visionary management, Camford Publishing Ltd. 354-358.

25)Michael H. McGivern and Steven J. Tvorik (1998), Vision driven organizations: measurement techniques for group classification, Management Decision. 242-252.

26)John Nicholls (1994), The heart head and hands of transforming leadership, Leadership & Organization Development Journal. 9-12.

27)John S. Leipzig (2004), Modeling the communication dynamics of uninspired leadership, Corporate communications: an international journal. 128-134.

28) Nurdan Ozaralli (2002), Effects of transformational leadership on empowerment and team effectiveness, Leadership and Organizational development Journal. 335-342.

29)Andrew Napier (2007), Incentives—A vector of improved sales performance?, Emerald Backfiles. 37-42.

30)Melissa A. Williams, Timothy B. Michael and Edward R. Waller(2008), Managerial incentives and acquisitions: a survey of the literature, Managerial Finance. 329-339.

31)Scott Weaven and Lorelle Frazer (2006), Investment incentives for single and multiple unit franchisees, Qualitative Market Research: An international journal. 227-235.

32)Jochen Wirtz and Patricia Chew(2002), The effects of incentives, deal proneness, satisfaction and tie strength on word-of mouth behavior, International journal of service industry management. 143-153.

33)Brian Ferguson and Jennifer N. W. Lim (2001), Incentives and clinical governance, Money following quality, Journal of management in medicine. 470-482.

34)Per Christensen and Sussi Handberg(1996), Forces and incentives in the promotion of environmental protection and improved working conditions, Environmental management and health. 5-9.

35)L. Kazatzopoulos, C. Delakouridis, G. F. Marias and P. Georgiadis (2006), An incentive-based architecture to enable privacy in dynamic environments, Internet research. 172-182.

36)Charlie Weir and David Liang (1998), Management buy-outs: The impact of ownership changes on performance, Henry Stewart publications. 262-266.

37) Edwin Joetan and Brian H. Kliener (2004), Incentive Practices in the US Automobile Industry, Management research news. 50-54.

38) Bjorge Timenes Laugen, Nuran Acur, Harry Boer and Jan Frick (2005), Best manufacturing practices, What do the best performing companies do?. Journal of Production and operations management. 133-143.

39)H. James Harrington(1997), The fallacy of universal best practices, The TQM Magazine. 61-71.

40)Bjorn Andersen, Tom Fagerhaug, Stine Randmoel, Jurgen Schuldmaier, Johann Prenninger(1999), Benchmarking: Supply chain management best practices. Journal of Business and Industrial Marketing. 378-384.

41)Mustafa Ungan (2004), Factors affecting the adoption of manufacturing best practices, Benchmarking: An International Journal. 504-514.

42)Payam Hanafizadeh, Morteza Moosakhani and Javad Bakhshi (2009), Selecting the best strategic practices for business process redesign. Business Process Management Journal. 609-619.

43)Robin Mann, Oludotun Adebanjo and Dennis Kehoe (1998), Best practices in the food and drinks industry. Benchmarking for Quality Management & Technology. 187-192.

44)S. Limam Mansar and H. A. Reijers (2007), Best practices in business process redesign: use and impact, Business Process Management Journal. 197-207.

45)Todd A. Boyle(2004), Towards best management practices for implementing manufacturing flexibility. Journal of Manufacturing Technology Management. 8-20.

46)Pui-Mun Lee (2002), Sustaining business excellence through a framework of best practices in TQM, The TQM Magazine. 143-147.

47)Richard Cuthbertson and Wojciech Piotrowicz (2008), Supply chain best practices-identification and categorization of measures and benefits. International Journal of Productivity & Performance management. 391-401.

48)Alberto Petroni and Maurizio Bevilacqua (2002), Identifying manufacturing flexibility best practices in small and medium enterprises. International journal of Operations and Production Management. 931-943.

49)R. Van Landeghem and K. Persoons (2001), Benchmarking of logistical operations based on a causal model. International Journal of Operations and Production management. 253-265.

50)Sergio Beretta, Andrea Dossi and Hugh Grove(1998), Methodological strategies for benchmarking accounting processes. Benchmarking for Quality Management & Technology. 167-170.

51)Ulf Johanson, Matti Skoog, Andreas Backlund and Roland Almqvist(2006), Balancing dilemmas of the balanced scorecard. Accounting, Auditing & Accountability Journal. 844-854.

52)Heinz Ahn(2005), How to individualize your balanced scorecard, Measuring Business Excellence. 6-11.

53)Sven C. Voelpel, Marius Leibold and Robert A. Eckhoff(2006), The tyranny of the Balanced Scorecard in the innovation economy. Journal of Intellectual Capital. 44-55.

54)Helen Atkinson (2006), Strategy implementation: a role for the balanced scorecard?, Management Decision. 1443-1456.

55)Vicky Rich(2007), Interpreting the balanced scorecard: an investigation into performance analysis and bias. Measuring business excellence. 6-9.

56) Juha Kettunen and Ismo Kantola (2005), Management information system based on the balanced scorecard, Campus-Wide Information system. 267-271.

57)jan Achterbergh, Robert Beeres and Dirk Vriens(2003), Does the balanced scorecard support organizational viability?. Kybernetes.1389-1401.

58)Ian Cobbold, Gavin Lawrie and Khalil Issa (2004), Designing a strategic management system using the third-generation balanced scorecard, International Journal of Productivity and Performance management. 625-630.

59) Teresa Garcia-Valderrama, Eva Mulero-Mendigorri and Daniel Revuelta-Bordoy (2008), A Balanced Scorecard framework for R&D, European Journal of Innovation Management. 242-260.

60) Ali Assiri, Mohammed Zairi and Riyad Eid (2006) How to profit from the balanced scorecard. Industrial management & data systems. 943-945.

61)Adrien Chia, Mark Goh and Sin-Hoon Hum (2009), Performance measurement in supply chain entities: balanced scorecard perspective, Benchmarking: An International Journal. 607-617.

62)Charalambos Spathis and Sylvia Constantinides (2004), Enterprise resource planning systems impact on accounting processes. Business Process Management Journal.235-244.

63)A.J.D. Lambert, M.H. Jansen and M.A.M. Splinter (2000), Environmental information systems based on enterprise resource planning, Integrated Manufacturing System. 106-110.

64)Rodney McAdam and Alan Galloway(2005), Enterprise resource planning and organizational innovation: a management perspective. Industrial Management & Data Systems. 282-287.

65)Carl Marnewick and Lessing Labuschagne (2005), A conceptual model for enterprise resource planning (ERP). Information management & computer security. 146-152.

66) Joseph R. Muscatello, Michael H. Small and injazz J. Chen (2003), Implementing enterprise resource planning (ERP) systems in small and midsize manufacturing firms. International journal of Operations & Production management. 852-568.

67)Majed Al-Mashari (2003), Enterprise resource planning (ERP) systems: a research agenda. Industrail Management & Data systems. 23-25.

68)Mike Kennerley and Andy Neely(2001), Enterprise resource planning: analyzing the impact. Integrated Manufacturing Systems. 105-110.

69)Satya P. Chattopadhyay (2001), Relationship Marketing in an enterprise resource planning environment. Marketing Intelligence & Planning. 137-139.

70)Fawzy Soliman, Stewart Clegg and Tarek Tantoush (2001), Critical success factors for integration of CAD/CAM systems with ERP systems. International Journal of Operations & Production Management. 611-626.

71)Thomas H. Davenport and Jeffrey D. Brooks (2004), Enterprise systems and the supply chain. Journal of Enterprise Information management. 9-15.

72) Injazz J. Chen (2001), Planning for ERP systems: analysis and future trend. Business Process Management Journal. 376-380.

73)Raymond F. Boykin and Wm. Benjamin Martz Jr (2004), The integration of ERP into a logistics curriculum: applying a systems approach. Journal of Enterprise Information Management. 46-51.

74)Michael D. Okrent and Robert J. Vokurka (2004), Process mapping in successful ERP implementations. Industrial management and Data Systems. 638-642.

75) Richard Fulford and Peter E.D. Love(2004), Propagation of an alternative enterprise service application adoption model. Industrial management & data systems. 451-455.

76)S. Maguire, S.C.L. Koh and A. Magrys (2007), The adoption of e-business and knowledge management in SMEs. Benchmarking: An international journal. 38-52.

77)vachara Peansupap and Derek Walter (2005), Factors affecting ICT diffusion. Engineering, construction and architectural management. 22-34.

78)Remko Van Hoek (2001), E-supply chains—virtually non-existing. Supply chain management: An international journal. 22-25.

79)Mairead Brady, Martin R. Fellenz and Richard Brookes (2008), Researching the role of information and communication technology ICT in contemporary marketing services. Journal of Business& Industrial Marketing. 109,110,111.

80) Oystein Moen, Tage Koed Madsen and Arild Aspelund (2008), The importance of the internet in international business-to-business markets. International marketing review. 489-494.

81)Marijn Janssen (2004), Insights from the introduction of a supply chain co-ordinator. Business Process Management Journal. 301-306.

82)Paul Hong, Be-Boong Kwon and James Jungbae Roh (2009), Implementation of strategic green orientation in supply chain. European Journal of Innovation Management. 514-516.

83) Bulent Sezen (2008) Relative effects of design, integration and information sharing on supply chain performance. Supply chain management: An international journal. 334-337.

84)Thomas H. Davenport and Jeffrey D. Brooks(2004), Enterprise systems and the supply chain. Journal of enterprise information management. 8-12.

85)Soo Wook Kim (2006), Effects of supply chain management practices, integration and competition capability on performance. Supply chain management: An international journal. 241-245.

86)Mickey Howard and Brian Squire (2007), Modularization and the impact on supply relationships. International journal of Operations & Production management. 1195-1206.

87)R.R. Bales, R.S. Maull and Z. Radnor (2004), The development of supply chain management within the aerospace manufacturing sector. Supply chain management: An international journal. 250-253.

88) Andrew R.J. Dainty, Geoffrey H. Briscoe and Sarah J. Millett (2001), New perspectives on construction supply chain integration. Supply chain management: An international journal. 164-167.

89)John Henneberry (2007), Interrelationships between manufacturing firms and accommodation in the rented sector: Theoretical perspectives. Property management. 213,214.

90)Wann-Yih Wu, Chwan-Yi Chiang and Jeng-Sin Jiang (2002), Interrelationships between TMT management styles and organizational innovation. Industrial Management & Data Systems. 172-177.

91)Ann P. Young (2000), I'm just me. Journal of Organizational Change. 377-379.

92)Rumy Husan (1997), The continuing importance of economies of scale in the automotive industry. European Business Review. 39.

93)Roger Hallowell (1999), Exploratory research: consolidations and economies of scope. International journal of service industry management. 359-364.

94)William J. Bratton, Robert J. Bennett and Paul J.A. Robson (2003), Critical mass and economies of scale in the supply of services by business support organizations. Journal of services marketing. 733-737.

95)Don Peppers and Martha Rogers (1995), A new marketing paradigm: share of customer, not market share. Managing service quality. 49,50.

96)Kalevi Kylaheiko and Jaana Sandstrom (2007), Strategic options-based framework for management of dynamic capabilities in manufacturing firms. Journal of manufacturing technology management. 967-980.

97)Stuart Hannabuss (1996), Linkage analysis: its use with staff and information. Library review. 8-14.

98)Carolan McLarney (2001), Strategic planning-effectiveness-environment linkage: a case study. Management Decision. 809-814.

99)Jason Hurwitz, Stephen Lines, Bill Montgomery and Jeffrey Schmidt (2002), The linkage between management practices, intangibles performance and stock returns. Journal of Intellectual Capital. 52-59.

100)Peggy A. Golden, Denise Johnson and Jerald R. Smith (1995), Testing the environment-strategy linkage in transition economies: A study of Russian managers. The international journal of organizational analysis.285-298.

101) Chang Won Lee, Ik-Whan G. Kwon and Dennis Severance (2007), Relationship between supply chain performance and degree of linkage

among supplier, internal integration, and customer. Supply chain management: An international journal. 444-450.

102)Jay S. Kim and Peter Arnold (1996), Operationalizing manufacturing strategy: An exploratory study of constructs and linkage. International journal of Operations & production management. 50-64.

103)Robert D. Klassen (2000), Exploring the linkage between investment in manufacturing and environmental technologies. International journal of Operations & Production management. 129-135.

104)Yongtae Park and Seonwoo Kim (2005), Linkage between knowledge management and R&D management. Journal of Knowledge management. 34-39.

105)Raffaella Cagliano, Frederico Caniato and Gianluca Spina (2006), The linkage between supply chain integration and manufacturing improvement programmes. Journal of Operations & Production management. 284, 286, 290.

106)Su-Chao Chang and Ming-Shing Lee (2008), The linkage between knowledge accumulation capability and organizational innovation. Journal of knowledge management. 3,4, 10-15.

107)Michiya Morita and E. James Flynn (1997), The linkage among management systems, practices and behavior in successful manufacturing strategy. International Journal of Operations & Production management. 968, 975-985.

108)James E. Ricks(1997), Electronically developed theory and procedure for distribution channel management via electronic data interchange linkage. Logistics information management. 22,24, 25-27.

109)Ralph A. Oliva (2006), The three key linkages: improving the connections between marketing and sales. Journal of Business & Industrial Marketing.395, 396, 397.

110)J. Stephen Town (2000), Performance or measurement?. Performance measurement and metrics. 43, 45, 50-54.

111)Andrew Booth (2006), Counting what counts: performance measurementandevidence-basedpractice.Performancemeasurement and metrics. 64,65-71.

112)Alaa M. Ghalayini and James S. Noble (1996), The changing basis of performance measurement. International journal of Operations & Production management. 65,66,70, 72-76.

113)Matthias Elg (2007), The process of constructing performance measurement. The TQM Magazine. 219-227.

114)Patrick Hoverstadt, Ian Kendrick and Steve Morlidge(2007), Viability as a basis for performance measurement. Measuring business excellence. 27-32.

115)Andy Neely (1999), The performance measurement revolution: why now and what next?. International journal of Operations & Production management. 207-222.

116)A.M.Ahmed(2001),Virtualintegratedperformancemeasurement. International journal of quality & reliability management. 415, 416, 417.

117)Arie Halachmi (2002), Performance measurement and government productivity. Work study. 63,64-68.

118)Umit S. Bititci, Allan S. Carrie and Liam Mc Devitt (1997), Integrated performance measurement systems: a development guide. International journal of Operations & Production management. 523-531.

119)Umit S. Bititci, Allan S. Carrie and Liam McDevitt (1997), Integrated performance measurement systems: an audit and development guide. The TQM Magazine. 46-52.

120)Hans de Bruijn (2002), Performance measurement in the public sector: strategies to cope with the risks of performance measurement. The International journal of Public Sector management. 579, 580, 582-591.

121)Paul Rouse and Martin Putterill (2003), The integral framework for performance measurement. Management decision. 791-805.

122)Jill MacBryde and Kepa Mendibil (2003), designing performance measurement systems teams: theory and practice. Management Decision. 722-733.

123)Johan C. van Schalkwyk (1998), Total quality management and the performance measurement barrier. The TQM Magazine. 125-129.

124)Diana Woodburn (2004), Engaging marketing in performance measurement. Measuring business excellence. 63-72.

125)Mike Kennerley and Andy Neely (2002), A framework of the factors affecting the evolution of performance measurement systems. International journal of Operations & Production management. 1223-1241.

126)Ian Robson(2004), From process measurement to performance improvement. Business Process management journal.511-519.

127)Mihalis Giannakis (2007), Performance measurement of supplier relationships. Supply chain management: An international journal, 404-406.

128)R.C. Barker (1995), Financial performance measurement: not a total solution. Management Decision. 34-37.

129)Prabir K. Bagchi (1995), Role of benchmarking as a competitive strategy: the logistics experience. International journal of physical distribution and logistics management. 4-13.

130)George Panagiotou (2007), Reference theory: strategic groups and competitive benchmarking. Management Decision. 1595-1605.

131)Prabir K. Bagchi (1997), Logistics benchmarking as a competitive strategy: some insights. Logistics information management. 28-33.

132)Scott G. Dacko (2000), Benchmarking competitive responses to pioneering new product introductions. Benchmarking: An international journal. 325-336.

133)Elaine Monkhouse (1995), The role of competitive benchmarking in small to medium-sized enterprises. Benchmarking for Quality Management & Technology. 42-46.

134)Stanley E. Fawcett and M. Bixby Cooper (2001), Process integration for competitive success. Benchmarking: An international Journal. 397-408.

135)Mohamed Zairi and Rob Hutton (1995), Benchmarking: a process-driven tool for quality improvement. The TQM magazine.35-38.

136)Alberto G. Canen and Geoff H. Williamson (1996), Facility layout overview: towards competitive advantage. Facilities. 6,7.

137)Hsiu Li Chen (2001), Benchmarking and quality improvement. International journal of Quality & Reliability management. 758-765.

138)Hsiu-Li Chen (2005), A competence-based strategic management model factoring in key success factors and benchmarking. Benchmarking: An international journal. 366-374.

139)Jose Maria Viedma Marti (2004), Strategic knowledge benchmarking system (SKBS): a knowledge-based strategic information system for firms. Journal of knowledge management. 32-46.

140) David Wainwright, Gill Green, Ed Mitchell and David Yarrow(2005), Towards a framework for benchmarking ICT practice, competence and

performance in small firms. Performance measurement and metrics: An international journal for library and information services.44-50.

141)Rodney McAdam, Shirley Ann Hazlett and Karen Anderson-Gillespie(2008), Developing a conceptual model of lead performance measurement and benchmarking. International journal of Operations & Production management.1160-1175.

142)Mario Binder, Ben Clegg and Wolfgang Egel-Hess (2006), Achieving internal process benchmarking: guidance from BASF. Benchmarking: An international journal. 664-680.

143)Andrea Rangone (1996), An analytical hierarchy process framework for comparing the overall performance of manufacturing firms. International journal of Operations & Production management. 105-114.

144)Fariborz Y. Partovi, Jonathan Burton and Avijit Banerjee(1989), Application of Analytical Hierarchy Process in Operations Management. Emerald Backfiles. 6-15.

145)Chin-Tsai Lin and Pi-Fang Hsu (2003), Adopting an analytic hierarchy process to select Internet advertising networks. Marketing Intelligence & Planning. 184-189.

146)Godwin G. Udo (2000), Using analytic hierarchy process to analyze the information technology outsourcing decision. Industrial Management & Data Systems. 421-426.

147)Li Yulong, Wu Xiande and Li Zhongfu (2008), Safety risk assessment on communication system based on satellite constellations with the analytic hierarchy process. Aircraft engineering and aerospace technology: An international journal. 595-602.

148)Selcuk Percin (2006), An application of the integrated AHP-PGP model in supplier selection. Measuring Business Excellence. 34-45.

149)Mark Davies (2001), Adaptive AHP: a review of marketing applications with extensions. European journal of marketing. 876-882.

150) Yee-Ming Chen and Pei-Ni Huang (2007), Bi-negotiation integrated AHP in suppliers selection. Benchmarking: An International journal. 575-584.

151)Paolo Taticchi, Flavio Tonelli and Luca Cagnazzo(2009), A decomposition and hierarchical approach for business performance measurement and management. Measuring Business Excellence. 47-54.

152)Peter L. Danner (2007), Exchange value in the value hierarchy. Emerald Backfiles. 70-79.

153)Shouhong Wang (2001), Designing information systems for electronic commerce. Industrial Management & Data Systems. 304-309.

154)Charles V. Trappey and Amy J.C. Trappey (2001), Electronic commerce in Greater China. Industrial management & data systems. 201-207.

155)Susan Oliver (1997), A model for the future of electronic commerce. Information management & computer security. 166-169.

156)Stephen Wilson (1997), Certificates and trust in electronic commerce. Information management & computer security. 175-180.

157)Kevin F. McCrohan (2003), Facing the threats to electronic commerce. Journal of Business & Industrial marketing. 136, 137.

158)Niklas Aldin, Per-Olof Brehmer and Anders Johansson (2004), Business development with electronic commerce: refinement and positioning. Business Process Management Journal. 44-56.

159)K. Ruikar, C.J. Anumba and P.M. Carrillo (2003), Reenginnering construction business processes through electronic commerce. The TQM Magazine. 197-209.

160)Holmes Miller (2005), Information quality and market share in electronic commerce. Journal of services marketing. 93-99.

161)Necmi Karagozoglu and Martin Lindell (2004), Electronic commerce strategy, operations, and performance in small and medium-sized industries. Journal of small business & industrial development. 290-299.

162)Benjamin P.C. Yen and Elsie O.S. Ng (2003), The migration of electronic commerce (EC): from planning to assessing the impact of EC on supply chain. Management Decision. 656-662.

163)Ramnath K. Chellappa and Paul A. Pavlou (2002), Perceived information security, financial liability and consumer trust in electronic commerce transactions. Logistics Information management. 358-366.

164)Jari Salo(2007), Business relationships redesign with electronic commerce tools. Business Process Management Journal. 488-499.

165)Sean Xin Xu, Xu Yan and Xiaona Zheng (2008), communication platforms in electronic commerce: a three-dimension analysis. Emerald Group Publishing Limited. 50-54.

166)Panagiota Papadopoulou, Andreas Andreou, Panagiotis Kanellis and Drakoulis Martakos (2001), Trust and relationship building in electronic commerce. Internet Research: Electronic Networking applications and Policy. 322-329.

167)Chia Chi Lin (2003), A critical appraisal of customer satisfaction and e-commerce. Managerial auditing journal. 202-209.

168)Dimitris Kardaras and Eleutherios Papathanassiou (2000), The development of B2C e-commerce in Greece: Current situation and future potential. Internet research: Electronic networking applications and policy. 284-292.

169) Urban Ljungquist(2008), Specification of core competence and associated components. European Business review. 74-87.

170)Yu-fen Chen and Tsui-chih Wu (2007), An empirical analysis of core competence for high-tech firms and traditional manufacturers. Journal of Management Development. 160-166.

171)Brian Leavy (2003), Assessing your strategic alternatives from both a market position and core competence perspective. Strategy & leadership. 29-34.

172)G.J. Bergenhenegouwen, H.F.K. ten Horn and E.A.M. Mooljman (1996), Competence development-a challenge from HRM professionals: core competences of organizations as guidelines for the development of employees. Journal of European Industrial Training. 29-33.

173)Amy V. Snyder and H. William Ebeling Jr (2007), Targeting a Company's Real Core Competencies. Emerald Backfiles. 27-30.

174)Bob Mansfield (2003), Competence in transition. Journal of European Industrial Training. 297-307.

175)Anders Pehrsson (2004), Strategy competence: a successful approach to international market entry. Management Decision. 758-765.

176) Daniel Vloeberghs and Liselore Berghman (2003), Towards an effectiveness model of development centres. Journal of Managerial Psychology. 511-532.

177)Martti Tapio Lindman (2007), Remarks on the quality of the construction of business concepts. European Business Review. 196-111.

178)J.C. Baker, J. Mapes, C.C. New, M. Szwejczewski (1997), A hierarchical model of business competence. Integrated Manufacturing Systems. 266-271.

179)Bjarne Jensen and Hanne Harmsen (2001), Implementation of success factors in new product development-the missing links?. European Journal of Innovation Management. 37-50.

180)Philip R. Lindsay and Roger Stuart (1997), reconstructing competence. Journal of European industrial training. 328-330.

181)Jay S. Kim and Peter Arnold (1992), Manufacturing competence and Business Performance: A framework and Empirical Analysis. International Journal of Operations & Production management. 4-17.

182)John Mills, Ken Platts and Mike Bourne (2003), Competence and resource architecture. International Journal of Operations & Production Management. 977-991.

183)Anders Ortenblad (2001), On differences between organizational learning and learning organization. The learning organization. 125-130.

184)Thomas Garavan (1997), The learning organization: a review and evaluation. The learning organization. 19-25.

185)Phillip C. Wright and Monica Belcourt (1995), Down in the trenches: learning in a learning organization. The learning organization. 35, 36.

186)Colin J. Coulson-Thomas (1996), BPR and the learning organization. The learning organization. 16-20.

187)Robert Kay and Richard Bawden (1995), Learning to be systemic: some reflections from a learning organization. The learning organization. 22-24.

188)David Battersby (1999), The learning organization and CPE: some philosophical considerations. The learning organization. 58-60.

189)Steven H. Appelbaum and Lars Goransson (1997), Transformational and adaptive learning within the learning organization: a framework for research and application. The learning organization. 115-126.

190)Stephen A.W. Drew and Peter A.C. Smith (1995), The learning organization: change proofing and strategy. The learning organization. 4-9.

191)Jon-Chao Hong and Chia-Ling Kuo (1999), Knowledge management in the learning organization. The Leadership & Organization Development Journal. 207-211.

192)David Limerick, Ron Passfield and Bert Cunnington (1994), Transformational Change: Towards an Action Learning Organization. The Learning Organization. 31-37.

193) Myra Hodgkinson (2000), managerial perceptions of barriers to becoming a learning organization. The learning organization. 157-164.

194)Lars Steiner (1996), Organizational dilemmas as barriers to learning. The learning organization. 193-199.

195)Charles S. Englehardt and Peter R. Simmons (2002), Creating an organizational space for learning. The learning organization. 39-45.

196)Thomas Robinson, Barry Clemson and Charles Keating (1997), Development of high performance organizational learning units. The learning organization. 228-234.

197) William D. Hitt (1996), the learning organization: some reflections on organizational renewal. Employee Counselling Today. 16-25.

198)Frances M. Hill (1996), Organizational learning for TQM through quality circles. The TQM Magazine. 53-57.

199)Marie-Helene Abel (2008), Competencies management and learning organizational memory. Journal of knowledge management. 15-29.

200)Svetlana Cicmil (1997), Achieving completeness through TQ principles and organizational learning. The learning organization. 30-34.

201)Dexter Dunply, Dennis Turner and Michael Crawford (1997), Organizational learning as the creation of corporate competencies. Journal of Management Development. 232-240.

202)Ivo de Loo (2002), The troublesome relationship between action learning and organizational growth. Journal of Workplace learning. 245-253.

203)Petra C. de Weerd-nederhof, Barnice J. Pacitti, Jorge F. da Silva Gomes and Alan W. Pearson (2003), Tools for the improvement of organizational learning processes in innovation. Journal of workplace learning. 320-330.

204)Cameron M. Ford and dt Ogilvie (1996), The role of creative action in organizational learning and change. Journal of organizational change. 54-59.

205)Peter Trim and Yang-Im Lee (2007), Placing organizational learning in the context of strategic management. Business Strategy series. 335-341.

206)Y.L. Jack Lam(2001), Toward reconceptualizing organizational learning: a multidimensional interpretation. The international journal of educational management. 212-219.

207)David Oliver and Claus Jacobs (2007), developing guiding principles: an organizational learning perspective. Journal of Organizational Change management. 814-825.

208)Stephen A.W. Drew and Peter A.C. Smith (1995), The new logistics management: transformation through organizational learning. Logistics Information management. 24-33.

209)Ricardo Chiva-Gomez, Cesar Camison-Zornoza and Rafael Lapiedra-Alcami (2003), Organizational learning and product design management: towards a theoretical model. The learning organization. 167-182.

210)Jane C. Linder (2004), Outsourcing as a strategy for driving transformation. Strategy & leadership. 26-31.

211)Marie L. Thorne (2000), Interpreting corporate transformation through failure. Management Decision. 305-314.

212)Umit S. Bititci (2007), An executives guide to business transformation. Business Strategy Series. 203-213.

213)Andy Adcroft, Robert Willis and Jeff Hurst (2008), A new model for managing change: the holistic view. Journal of business strategy. 40-45.

214) Jane Fiona Cumming, Neela Bettridge and Paul Toyne (2005), responding to global business critical issues. Corporate Governance. 42-48.

215)Evangelina D. Fassoula (2006), Transforming the supply chain. Journal of manufacturing technology management. 848-858.

216)Carl-Henric Nilsson and hakan Nordahl (1995), making manufacturing flexibility operational-part 1: a framework. Integrated manufacturing systems. 5-10.

217)Michael A. Lewis (2003), Analyzing organizational competence: implications for management of operations. International journal of operations & production management. 738-751.

218)Angeliki Poulymenakou and Loukas Tsironis (2003), Quality and electronic commerce: a partnership for growth. The TQM Magazine. 139-147.

219)Christian N. Madu (2004), Strategic value of reliability and maintainability management. International journal of quality and reliability management. 318-326.

www.ingramcontent.com/pod-product-compliance
Lightning Source LLC
Chambersburg PA
CBHW031830170526
45157CB00001B/255